PRICE THIRTY-FIVE CENTS.

SF523
M588

METCALF'S KEY

TO

BEE-KEEPING.

NEW-YORK:

C. M. SAXTON, AGRICULTURAL BOOK PUBLISHER.

1862.

THE MICHIGAN BEE-HIVES.

————o————

The inventor of these Hives has obtained two Patents, from which we quote as follows :

U. S. Patent Office.

Letters Patent, No. 1,948, whole No. 32,952, dated July 30, 1861.

The nature of my invention consists in the peculiar construction and employment of the various parts which compose the hive, and also in the particular mode of producing artificial swarming, to which purpose the parts hereinafter set forth are merely incidental. [See Fig. 19, p. 36.

Claims.—1st The employment of a revolving bee-hive, so arranged that artificial swarming may be produced substantially in the manner specified.

2nd. I claim the employment of the movable frames, D. D . provided with cylinder M, when used in connection with a revolving bee-hive, in the manner and for the purpose set forth.

————————

U. S. Patent Office.

Letters Patent, whole No. 34,157, dated Jan. 14, 1862.

Fig. 1 is a perspective view, showing the construction of the frame. [See Fig. 23, p. 47]

Fig. 3, side view of frames in hive. [See Fig. 22, p. 46]

These adjustable frames are intended to be used in the revolving bee-hive, patented to me July 30, 1861, or in any rectangular box hive, having a movable front.

Claim.—I claim constructing the top bar A, and side bars B, of adjustable frames for bee-hives, with the beveled ends *a*, *b*, in the manner described, when used in connection with a movable front, and in a rectangular box or hive.

————o————

ITALIAN BEES,

Put up securely, and forwarded by express (C. O. D.), and a safe delivery, and successful Italianization of the common hive guaranteed.

PRICE PER QUEEN, FIVE DOLLARS.

MARTIN METCALF,
Grand Rapids, Mich.

A KEY

TO

SUCCESSFUL BEE-KEEPING:

BEING A TREATISE ON THE

MOST PROFITABLE METHOD OF MANAGING BEES,

INCLUDING THE AUTHOR'S

NEW SYSTEM

OF

ARTIFICIAL SWARMING,

WHEREBY

ALL WATCHING FOR SWARMS DURING THE SWARMING
SEASON IS DONE AWAY WITH,

AND

All Loss by Flight to the Woods Prevented.

BY
MARTIN METCALF.

"Who guides the patient pilgrim to her cell?
Who bids her soul with conscious triumph swell?
With conscious truth retrace the mazy clue
Of various scents that charmed her as she flew?"

NEW-YORK:
C. M. SAXTON, AGRICULTURAL BOOK PUBLISHER.
1862.

E 1125

J. J. REED, Printer and Stereotyper,
43 Centre Street.

PREFACE.

THE author of this little book and originator of the system of Bee-Keeping herein advocated, has endeavored to present the results of many years of careful observation, experiments and study in the form of a practical treatise, so concise and yet so plain, that even the inexperienced may, with his book and his hive, enter upon the business of bee-keeping confident of success.

He takes pleasure in acknowledging that in his study of the Bee he has derived much valuable aid from the labors of Bevan, Huber, Huish, Miner, Taylor, Munn, Bruckish, Quinby, Langstroth, Harbison and others. Bringing their conflicting theories to the only sure test—the Bee-hive itself, so constructed as to expose to the eye its entire operations—he has, by careful observation at all hours and all seasons, and continued through many years, demonstrated the truth of some and the error of others. Thus, little by little, under his own eye, have the natural laws governing these wonderful little creatures, ar-

3

ranged themselves into a compact and beautiful system. His experiments were pursued simply as a pleasure, and his hive invented for his own private use : it was not until quite recently that he had any thought of offering to the public a new hive, or publishing a book on Bees. He now does it at the request of friends, and to supply what seems to be a public need.

He has intended to give due credit in his book.— Its designed limits and chief aim forbade extensive copying ; for he has purposely excluded all speculative hypotheses not yet brought from the field of experiment by repeated demonstrations, giving only the absolute and abundantly established truths on which the hive and system herein described rest. He does not expect, in a day, to convince all of the truth of every statement made in his book, nor of the entire practicability of his system. He asks for the former a careful study, and for the latter a fair trial—these granted, he has no fears for the result.

THE AUTHOR.

Grand Rapids, Michigan, April, 1862.

INTRODUCTION.

"But these pursuits will honeyed fragrance bring
Without the danger of a treacherous sting."

BEES suggest all that is beautiful, fragrant and delicious in the floral universe. Hence bee-keeping has been termed the "poetry of agriculture." A flower without a bee to sip its nectar and rolic in its pollen, hints too broadly the *quasi* bliss of "single blessedness." Types of toil, symbols of frugality, models of government—with Flora propitious, how extravagantly provident, and how cheerfully they fill our dish with a "Benjamin's mess" of their delicate fare.

Whether bees are partial to good society, or man appreciates bees as he himself becomes refined—their temples have always marked the *locum in quo* of the highest style of human culture. Bees navigated the Nile in its palmiest days, to gather luscious wealth from her blooming fields. Phœnicia, preceptor of the world, was graphically described to its "heirs of promise," as "a land flowing with milk and honey." Greece, "the land of scholars," had her Mount Hybla—"the empire of bees"—and Emelus of Corinth, in 741 B. C., devoted a poem to their praise. Rome's most elegant poet, Virgil, sang the bee in the noon of her splendor. The learned Germans, importing the bee from Italy, and copying

5

their hive from Greece, have in some cases a thousand colonies to the square mile, sustain a *Journal*
and Associations in their interest, and government
encourages their culture. Of the 202 species of the
Apis genus, Hardie, in his America, says there are
111 in England, where they receive deserved attention.*

---

* The following anecdote from the Mark Lane *Express*, is in
point :

A bishop was holding his first visitation of the clergy of his
diocese, in a town in one of the midland countries. Among
those assembled he soon discovered an old college acquaintance,
whom he had not seen for a great number of years, but whom
he greeted with all the warmth of a renewed friendship. On
comparing notes with his friend, the bishop learned with regret
that he was still a curate in a country village, at a stipend of
one hundred pounds a year, and that he had a wife and large
family to support. The worthy curate, however, invited the
bishop to spend a day with him before he left the neighborhood,
and the latter, not wishing to appear proud, accepted the invita-
tion. On reaching the parsonage, he was surprised to find his
friend's wife an elegant, well-dressed lady, who received him
without any of the embarrassment which a paucity of means
occasions in those who feel its pressure. The children, too,
were all well-dressed, and looked anything rather than as having suffered from the pinching pains of unappeased hunger. But
the good bishop's astonishment was still greater when he sat
down to partake of a repast worthy of the traditional and cus-
tomary fare of his order, and was invited to " take wine" of the
purest flavor and aroma with his fair and graceful hostess.
Knowing that his friend was originally a poor man, he consider-
ed that he must have received a fortune with his wife. After,
therefore, the latter and the children had withdrawn, the bishop

Bees came with the Puritan fathers to the New World in 1670, and have long since become one of our economic necessities. They followed their descendants to California in 1853 and subsequent years, whence come fabulous accounts of their prodigious thrift.

With staple tribute for our tables and dollars for our pockets, they pass out of the realm of fine

---

introduced the subject by expressing a fear that his friend had gone to an unusual and injurious expense to entertain him, and that it would entail privation upon him afterwards.

"Not at all," replied the curate. "I can well afford to entertain an old friend once in a while without any inconvenience."

"Then," rejoined the bishop, "I must congratulate you, I suppose, on having received a fortune with your good lady."

"You are wrong again, my lord," replied the poor curate. "I had not a shilling with my wife."

More mystified than ever, the bishop resumed :

"Then how is it possible for you to have those comforts around you that I see, out of a hundred a year ?"

"Oh, my lord, as to that, I am a large manufacturer as well as clergyman, and employ many thousands of operatives, which bring me in an excellent living. If you will walk with me to the back of the premises, I will show you them at work." He accordingly took him into the garden at the back of the house, and there was a splendid apiary, with a large number of bee-hives, the source of the curate's prosperity.

The bishop never forgot the circumstance, nor did he ever fail to make use of it as an argument ; for when he afterwards heard some poor curate complain of the scantiness of his income, he would cut the matter short by exclaiming :

"There, there, let's have no more grumbling. Keep bees, like Mr. —— ; keep bees, keep bees !"

arts, and assume a commanding place in the sober economies of life. Safer than bank, railroad, or government stocks, and returning annually, with moderate attention, at least one hundred per cent. net on the capital invested, they might well claim a nitch in Wall-street. Adapting themselves to every section of our country, they will for only a quiet nook in the yard, make us independent of sorghum, or the cane, by gathering and storing away in sealed cans ready for our use, the wasting sweets of garden, field and forest—pure, healthful, and tempered to the palate beyond the most exquisite culinary art.

" Eh ! oh !—I like the honey and admire the bees, but——"

Never mind ; that sting was not made for you, and will not be used against you when you learn to treat them properly. But if you persist in rudely disregarding their comfort and their rights, returning every friendly salute with a blow, disturbing, crushing them *ad libitum*—even *murdering* whole colonies for their stores when you can get them much easier without—you deserve to ——become more civil and consult your own interest if not theirs, for the country needs a million bees where it has but one. Their sting was made for robbers, and for their insidious enemy, the moth, and will guard their stores and yours if you will give them a fair chance.

---

*The moth breaks through at night sometimes, and steals after a very singular manner--namely, by giving. She lays her eggs in the hive, cuckoo fashion, and their voracious *larvæ* de-

With modern improvements, bee-keeping is made a safe, sure, *humane* and pleasant business. After the hives are made—artificially swarming hives of course—the ladies can swarm the bees and do all else that needs to be done in the apiary. With the modesty of real worth, and generous often to a fault, these wonderful little creatures commend themselves alike to the naturalist, the amateur, the moralist, and the " solid" man.   ****.

---

vour the honey. [Beebread and wax ?] No wonder, then, that the bees fear moths—*timent Danaos et dona ferentes*, [" They fear the Greeks even bearing gifts"] a scholarly quotation which we have not often so good an opportunity of introducing.— *Chambers' Edinburgh Journal.*

SUCCESSFUL BEE-KEEPING.

A SWARM OF BEES,—OF WHAT IT CONSISTS.

A prosperous colony of bees, in the midst of the swarming season, consists of a single queen, from fifty to five hundred drones, and twenty thousand to fifty thousand workers.

The Queen, if fertile, is the only perfectly developed female in the hive. She is the mother of the whole colony, laying all the eggs, producing queens, workers, and drones. She has six abdominal rings, while workers and drones have only five; and she has over the thorax longitudinally a clearly defined line—a feature which the writer has never seen noticed by any author.

1.—Queen.

Drones are the male bees, and are of no use whatever, except to impregnate the young queens. This is done upon the wing, in the air, and within the first twenty-one days of the queen's existence.

2.—Drone.

The workers are undeveloped females; and, unlike the queen, they are incapable of fertilization, by copulation with the drones; yet are capable, under certain circumstances, of depositing eggs which will produce perfect drones, differing in no respect from those

3.—Worker.

hatched from eggs laid by either a fertile or an unfertile queen. The average time of maturity of a *worker* bee from the egg is twenty-one days ; drone twenty-four, queen about eighteen.

Having thus glanced at the different classes of bees found in a hive, in its prosperous condition, we will now take a look at the interior of the honied temple, witnessing their labors, noting the manner of their development, and the more prominent characteristics of each class.

THE QUEEN.

"First of the throng, and foremost of the whole,
One stands confessed the sovereign and the soul."

The queen lays all the eggs, and continues this labor the whole year round, the least brood, in our climate, being found in December. In January the brood increases, and more and more rapidly as spring approaches. The greatest amount of brood is found in June or July, or exactly at that point of time when the old queen leads out the first swarm of the current year. Indeed, the lack of cells in which to deposit her eggs, appears to be one of the causes of the issue of the first swarm, for it is found that the queen becomes much agitated on finding the breeding cells all occupied, even though the hive be not half full of comb. The workers, however, instinctively prepare for the migration of the mother, by providing cells for the rearing of young queens to supply her place. These cells are constructed on the extreme edges of the combs, and are in appearance not very unlike small acorn cups, with their open ends downward, attached at the base or upper end to the worker cells, which are nearly horizontal. Whether the queen herself deposits the eggs directly in the royal cradles, is not known. I believe she has never been seen to do so. I have myself many times kept watch to see how, and by whom this is done, without determining it. But be this as it may, whether laid there by the queen, or carried thither by the workers, as some suppose, that they are found there, and the embryo queens capped over while the old queen yet remains in the hive, is conclusively established. Before this time, all the breeding cells are occupied, either with honey, bee-bread, or brood, and the queen, becoming restless, perhaps from this cause, day by day moves more and more rapidly over the combs. The workers, too, partaking of the excitement, at first a few, their numbers gradually increasing, are seen running rapidly over the combs, striking their antennæ upon each other, until finally, as if by preconcert, rushing to the honey cells, unclosing many that have been sealed over, they fill themselves with their precious stores as eagerly as if they momentarily expected a writ of ejectment to be served upon them, and this was their last chance. During this scene *within*, all is unusually quiet *without* the hive ; while such bees as have been lying about the entrance, driven thither by the great heat or numbers within, now gradually wend their way back, whether to take their places in the old or new colony we will not stop to inquire. After each bee has taken on as large a load as it can carry, at the "appointed time, wind and weather permitting," they rush, pell-mell, from the hive, pouring out, and off the alighting board, like running water, many a greedy fellow falling to the ground from mere inability to fly with its too great self-imposed burden.

Swarming. — The queen usually leads out the first swarm from the second to the fourth day after the workers have commenced nursing the embryo queens. If the weather should then prove unfavorable for swarming, the young queens are destroyed. On the approach of a more congenial season, the work of queen rearing begins anew, to be repeated, it may be, again and again, and not unfrequently without swarming at all during the whole summer. From the ninth to the fourteenth day after the issue of the first swarm, the young queens will emerge from their cells, when, if the bees are still numerous, weather propitious, and the honey-yielding blossoms plenty, second, and third, or *after* swarms may be expected. The bees having previously divided off into as many "squads" as there are queens maturing, the first queen that issues from her cell generally leads out a *second* swarm, and, in one, two, or three days at farthest, if bees are still in considerable numbers, with other circumstances favorable, a third, fourth, and sometimes a fifth swarm issues. If the bees are not numerous at the time of the hatching of the first of the young queens, she is allowed her liberty, and will at once seek out and sting her rivals in their cells. If the hive be well filled with bees from the now rapidly maturing brood, those of each queen cluster will stand guard and prevent the queen from accomplishing her purpose, and others are allowed to hatch. Now may be heard the challenge of the queens to mortal combat, for one only can become fertile and remain in the hive. The "piping of the queens" may always be heard the morning or evening preceding the issue of all swarms *after the first*. If it be not heard by the fourteenth day after the issue of the *first*, no *after* swarm need be expected, and swarming is done with *that hive* for a period of forty days.

The first fair day after the hiving of a second, or after swarm, if a close watch be kept just before and during the flight of the drones—from 12 M. to 4 P. M.—the young queen may be seen to issue from the hive, and, taking wing, fly off into the air. This

4.—Unfertile. is called her " bridal trip," and is sometimes repeated every day, and several times a day, for many days. When successful, she returns to remain permanently in the hive, fully competent to supply eggs for the whole colony ; and the bees, at once clustering about her as they have never before done, show that *they* are conscious of the fact and recognize her sovereignty. The

practised eye of the experienced apiarian will at once detect this change in the affections of the bees, as well as a decided difference in the form and size of the queen herself the moment she touches the alighting-board. Before her departure they paid but little attention to her, running over her as freely as

5.—Fertile. over each other ; after her return they treat her with the utmost deference—never clamber over her, always clear the way as she approaches, and with their antennæ wave a " God save the queen" as she passes ; her every wish is anticipated, and her pleasure served with alacrity.

So, also, a day or two later, a young queen flies *from the parent stock* for the same purpose ; and if, as is too often the case, these hives stand in close proximity to others similar in shape, size, and appearance, a mistake is sometimes made by the returning queen in seeking entrance at a neighboring hive containing a fertile queen. Instant death awaits her here, and the future

destruction of her colony is rendered inevitable (unless
the remedy be applied), for the reason that no eggs are
now in the hive from which a queen can be reared, the
old and *fertile* queen having left two weeks before. A
month or two later in the season, a wonderful display of
drones takes place every afternoon ; and toward autumn,
if perchance the bee-keeper (?) happens to "heft" the
hive to ascertain how the honey harvest progresses, he
finds it wofully deficient in weight, that most essential
requisite of a bee-hive at this season of the year. A
closer inspection discloses the fact that the interior of
the hive can boast of *more worms than bees*, if, indeed, some
neighboring robber bees have not already discovered its
condition and taken charge of the disconsolate orphans
and their precious stores ; while the *bee-destroyer* — for I
shall call no such person a bee-keeper — goes straight to
his neighbor and tells the old and oft-repeated story that
"*the robbers and moth have ruined his bees !*" All stuff ! Let
such person get a hive giving perfect control of all its
combs, and facilities for the ready inspection of its contents, and he may soon be convinced that the loss of his
bees is justly chargeable only to himself. Having thus
rapidly glanced at the prominent characteristics of the
queen, and incidentally seen something of the workers
also, we will now turn our attention to the

DRONES.

"These lazy fathers of the industrious hive."

They are the male bees, and are ordinarily the offspring
of the queen, although it sometimes happens, that in the
absence of the queen, workers are found laying eggs that
produce them. The drones are short lived, averaging but

about two months, even when not meeting with violent
death at the hands of the workers, or by reason of their
fulfilling the object of their existence, to wit, the impregnation of the young queen ; for the cohabiting drone immediately dies. When queen-rearing is done for the season, the workers fall upon, and destroy them all.

THE WORKERS.

"So work the honey-bees,
Creatures that by a rule in nature, teach
The art of order to a peopled kingdom."

These constitute the great bulk of every prosperous
colony of bees. It is by their labors that the rich stores
of honey are collected, the delicately wrought cells and
systematically constructed combs are made, their temple caulked and well plastered with propolis, the abundant collections of farina, or pollen, provided and stored
away for food, and every want of the constantly maturing broods supplied. A guard of workers also attends the
queen in her almost ceaseless rounds, feeding and watch-

ing over her with sleepless vigilance, prepar-
ing the tiny cells for the reception of her eggs,
capping over, at the proper time, these cradles for their young, with thin scales of wax,
elaborated from between their abdominal rings,
—in short, (except the work of depositing
eggs,) all the labor of the hive is carried on
Fig. 6.*   by them. When a young bee emerges from
the cell, and the queen, passing along, refuses to place
an egg therein, immediately, and without the least per-
ceptible exchange of word or look of authority, by hint
or deed, the industrious and provident workers set themselves about repairing and fitting it up for another ten-

*Abdomen of a worker magnified, showing scales of wax.

ant ; a new egg is soon laid, and at once the watchful
workers keep vigil there. Four to five days elapse, and

7.—Brood.

the tiny egg breaks into a shapeless
mass—save that something like an
embryo worm is revealed by the
microscope ; when worker after
worker is seen peeping into the cell,
nursing and providing for the needs
of its occupant. Scarce a moment
can pass, but some new comer deposits its apportioned
food. Night and day this work of watchfulness and care
continues without interruption, while generation after
generation come and go from year to year. No confusion, not the slightest discord, even for a moment, but
each, intent on its appropriate task, right gleefully plies
its willing hands and feet, and all to one continual work,
work, WORK ! And amidst all this multiform labor, *order
reigns*. If any cells are found of too great depth, in consequence of having been used for the storing of honey
for the previous winter, they are at once cut down to the
required length on the approach of the broods to their
vicinity. Mathematical precision is apparent in all the
household ! From whence ? What silent, magic influence pervades the honied dome, and guides its myriad
throng ? We will open the hive and take away the
queen. Here she is ! Now it is closed again, and while
I hold her in my hand here, let us keep a close watch
upon the movements in the interior of the hive, through
its glass sides. Where but a moment since all was order
and work, what have we now ? See ! The bees are running here and there. The tumult increases ! Fifteen minutes have scarcely elapsed, and tumultuous disorder has
broken out through all the camp. All appears alarm and

confusion, even the *lazy drones* are taking wing, and, like a flood, the impetuous workers chase each other from the hive. Mounting in spiral circles high in the air, they soon return to repeat the search for the lost one, over and over again.

We will return the queen, letting her go just at the entrance of the hive. How quickly she darts in ! and the bees rush back in eager haste to get inside, and know for themselves that indeed the lost one is found. Immediately a loud buzzing, which, once heard, can never be mistaken, is audible. It is the glad shout of the hundred sentinels at the door, with uplifted wings and eloquent words, as clearly and plainly as any man could speak it, of the " all's well !" of the watchful guard, and the wonderful little creatures ply their busy hands, and feet, and wings, as before.

Again let us remove the queen, but this time not to return her. The tumult again takes place, and again subsides. Keep a close watch—we have a good opportunity to observe them. A single card of comb between two clear, clean panes of glass exposes both of its sides, so that we may see every bee. An hour has past. The bees have become quite composed. And see ! there is a cluster gathering here, another there, and over on the other side one, two, or three more. The bees are destroying some of their young, and dragging them from their

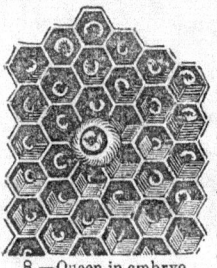

cells ! They have selected *one cell* in the centre of each of the clusters, which cell they are enlarging, most laboriously *pressing outward* its hexagonal sides, making it quite round. In a single day it has taken the appearance of an acorn cup, now beginning to curve downward. Three days more and it is capped over,

8.—Queen in embryo.

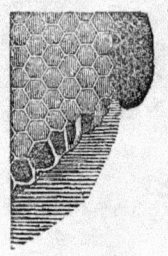

looking for all the world like a pea-nut depending perpendicularly from the comb, with its smaller end downward. Five days more, and the bees remove the wax covering from its tip, exposing a light brown silky substance, which is the cocoon spun by the young bee while in process of transformation from a worm to a winged insect. Two to three days more and *there emerges a bee which, but for our interference, would have been a worker, but which is now metamorphosed into a perfect*

9.—Capped over.

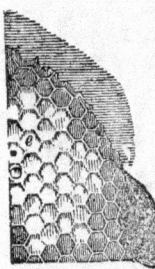

10.—Hatched.

queen! Almost immediately she seeks the rival royal cells, and, biting open their sides near the base, stings the luckless infants in their very cradles. The bees look quietly on until the deed of death is done, when they drag forth the lifeless bodies from the cells and from the hive ; unless, indeed, the hive be full of bees to overflowing, in which case the first emerging queen is prevented by the bees from destroying her rivals, and one issues with a young swarm. Thus, we can sometimes *compel a swarm* to issue even from a box hive, by removing the queen ; or, by introducing a queen cell so protected that the queen can not get to it.

It sometimes happens that the young queen fails to return to the hive—such a swarm becomes hopelessly queenless ; the bees seem utterly bereft of their faculties, refusing to work more than from hand to mouth, are irritable, quarrel with, and destroy each other.

11.—Queen cells destroyed.

" An angry hive of bees
That want their leader, scatter up and down,
And care not whom they sting."

If this occurs when the blossoms are yielding honey in profusion, a worker will be found to deposit eggs.

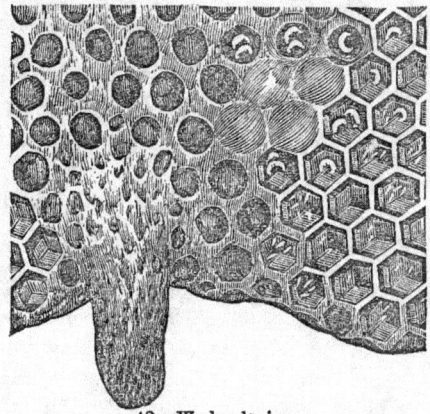

12.—Worker laying.

These eggs are laid regardless of order, scattered all over the combs, and frequently six or seven in a cell. They will hatch, the bees destroying all except one in each cell ; the remainder will mature *as drones only.*

Repeated attempts are made by the bees to rear queens from such eggs, which result in failure. They not unfrequently get so far as to cap over the cells and wait for the issue of the longed-for queen. But the hope is vain, for no *Royalty* is there, and the sleep of the grave*

* It will be observed as a singular fact, that while the eggs of both an *unfertile queen* and a *drone laying worker* will mature and hatch from either *drone* or *worker* cells, *they never mature as queens, nor hatch from the queen cell!* A perfect drone to all appearance will mature, but would seem to expire at the moment of hatching.

only rests within the *royal cradle* reared with so much so-
licitude and care, while the disconsolate, cheerless flock,
being rapidly decimated, hum listlessly in and out of the
hive with coarse, rough voices, ungainly carriage, and
murmurs plainly audible. Just place a queen among
them—in an instant their voices change. The well
known hum of joy is heard within, and answering voices
on the wing without, join in the chorus, while quick as
thought, away flies many a winged messenger, to distant
heath and forest flower, to return with bright golden pel-
lets of pollen, and heavily laden with delicious nectar to
proffer to their newly-found queen.

ABILITY TO SUPPLY THEMSELVES WITH A QUEEN.

The ability of a swarm of bees to supply themselves
with a queen from an egg which under other circumstances would have produced a worker, although not known
to the mass of bee-keepers, even of the present day, is not
a new discovery, but has been well known for many years,
and various methods of artificial multiplication of colonies
founded thereon have been proposed. Some of the most
plausible of these methods we purpose here to notice,
giving the claims of their inventors, and at the same time
pointing out as briefly as possible their practical work-
ings, while we shall also notice their defects and endeavor
to show just wherein they have been, to a greater or less
extent, failures, that the intelligent reader may contrast
these various methods with our own system, and judge for
himself of their comparative merits.

First, then, let us notice the *Piling*, or *"Nadir Hiving"*
system, so called because it consists in placing a series
of boxes, one upon another, leaving the bees to work

downward, taking from the top full ones either of honey-combs or bees, as wanted, supplying empty ones under-neath. Sometimes this seems to work, both swarms, when a division is thus made, prospering for a time ; at other times one or the other of the two portions proving a fail-ure altogether. It was not till after several years that this system, so fascinating in theory, was found to be worthless and the reasons pointed out. As this mode of multiplication sometimes *seems to work* and at other times to fail, and these failures result at different times from different causes, we shall be under the necessity of in-quiring what the bees of each division do under every phase of circumstances. Let three boxes, *A*, *B*, *C*, placed one above the other, represent one of these hives, *A* being at the top. In attempting to artificially swarm the bees with this hive, suppose *A* to be moved to a new location, and a new box, *D*, placed beneath *B C* — if the queen and most of the bees happen to be left in *B C*—what will be the result ? I venture to say, that in ninety-nine cases out of a hundred the bees will leave *A* and go back to *B C*, carrying the honey with them. Now suppose the queen to be taken in the box A, what will be the result ? In nine times out of ten, the establishment of a new colony in *A*, especially if a considerable number of bees and brood in the combs accompany the queen thither. But the greater portion of the swarm will return to the accus-tomed spot, *B C D*, which is sought to be prevented by removing *B C D* also a little distance. In either case considerable new comb will be built in *D* before the ma-turity of the young queen, the quantity being in propor-tion to the number of bees in the hive, not unfrequently *filling it up with worthless combs ;* since every colony of bees, while destitute of a queen, build *drone combs only.*

These drone cells number four to the inch each way, making sixteen on each side, or thirty-two cells to the

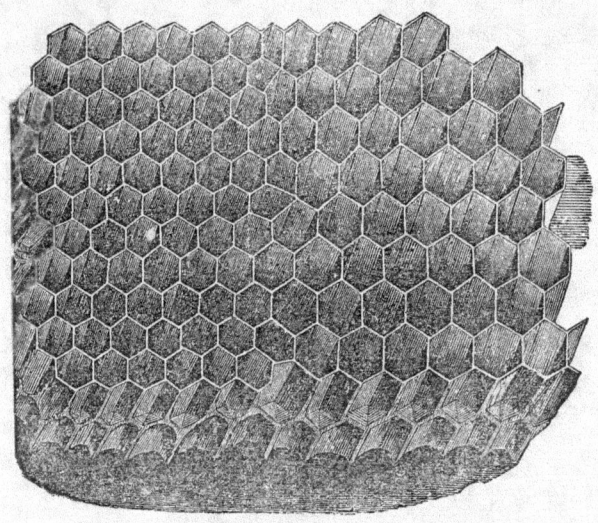

13.—Drone and Worker Comb, showing the transition from one to the other.

square inch of comb ; while the worker cells are five each way, or fifty to the square inch of comb. Let it be remembered that QUEENLESS COLONIES BUILD DRONE COMBS ONLY,* and that *worker* bees never issue from drone cells, and it will be evident why the " Piling" system is a fail-

* Except sometimes for about a day immediately following the removal of the queen.

ure. So also, as a general rule, bees construct store combs—deep cells—or large drone cells in all side or top boxes, or other apartments than that in which the queen and the broods are at the time.

The Dividing Hive was only another form of the same thing, and is open to the same objections and from precisely similar causes.

Another plan still, called the *Colonizing System*—and I speak of and expose the falacies of these various methods here because they are still adhered to by some who know no better, and shameless pretenders are to-day imposing upon public credulity by vending these exploded humbugs —is to make a box, and partition it off into two or more apartments having communications or openings between them, and also direct outlets to the fields. Into one of these a swarm of bees is put, it being intended that when the parent hive becomes filled, the bees, for want of room, shall pass through into the side apartments, gradually filling them with combs and brood. When this is done, the communication is to be cut off by means of a slide, and the part which is thus made queenless left to rear queens and colonize themselves. Like the methods before noticed, this, too, would *seem* to work sometimes, and why not always?

Because the system is based upon error and cannot possibly succeed, for the reason that not until the queen becomes crowded for cells in the main apartment in which to deposit her eggs will she pass into a side apartment. Before she passes into the side apartment, drone and store combs for the most part are built in that apartment ; and if during her absence from it and before it is occupied by brood, an attempt to swarm the bees be made by cutting off the communication, they will soon find it out and leave for home, for home is where the queen and the broods are, and there, too, the honey will go. If the separation happen to take place while the queen is in the side apartment, the thing may seem to work ; the parent stock prospering, but the new colony is pent up for breeding space for worker bees, and will never thrive.—By which I do not mean that it may not live, and struggle along for several years, *and be called a swarm of bees ;* but that it will not throw off good strong swarms, nor yield any considerable amount of surplus honey.

Indeed, all these hives have the same faults ; the greatest being the over-production of drone comb, thus contracting the breeding space for the queen till the season for drone rearing comes, when a flood of these appear, consuming all the surplus of the workers, now daily diminishing in numbers. Thus it is that the apiary soon "runs out," as the phrase is. But this is only the legitimate result of a system founded in error ; and it cannot be remedied even by the use of movable frames operated on the same erroneous principle, as many have sought to do. It is true, the *scientific bee-keeper may,* with considerable diligence and care, control drone breeding to a certain extent by the use of frames on the system above described ; yet he must be exceedingly careful that his queen-rearing swarms are not large and in possession of vacant space to be filled up with worthless combs. It is conceded that bees consume from twenty to twenty-five pounds of honey to elaborate one of wax.* Besides this, a swarm destitute of a queen is in an unnatural condition, and, however large, labors mainly to supply the present necessity, which is to rear a queen and drones to fecund-

---

* Whether it takes twenty-five pounds of honey to make one of wax, I am not prepared to prove ; probably it is near the truth.—*Quinby.*

ate her ; hence it is that no *worker* comb is built till a young queen hatches. That no stores of surplus honey are ordinarily gathered at such a time, one may satisfy himself by examining the honey boxes between the issue of a first and second swarm—the greatest amount being found about twenty-four hours before the issue of the first swarm. Again, a pint of bees will rear a queen in twelve days, and a bushel will do it no sooner.

The great secret of successful bee-keeping lies in keeping the stocks strong, and in swarming them artificially by a method that shall secure the construction of perfect worker combs in the greatest possible abundance, and save all the swarms without loss by flight to the woods. Such a method will also secure the greatest number of worker bees early in the season, and their greatest activity throughout the honey harvest. A safeguard against loss of swarms during our cold winters is also an essential element in a hive.

To combine these advantages, among others, into a practical system, has long taxed the ingenuity of intelligent bee-keepers. For the last thirty years there has been a marked advance in the right direction. A device, simple, cheap, and practicable, for obtaining control of the combs, has been one of the objects sought. Amateur apiarians placed a single card of comb in a thin case with glass sides, in order to observe the bees at work, and learn their habits. To insure the building of the comb in the right direction, Huber started it by fastening a small piece of comb to the ceiling ; he also combined eight of these cases or frames, hanging them by hinges, so that they would swing like a door, leaving out the glass sides, except the two outer ones, making the " *Leaf Hive* ;" which was invented more than sixty years ago. Bevan, Golding, Huish, Dzierzon, and others, used " *bars*," placed across the hives in rabbets, to which the bees built their combs. The side attachments had to be removed by cutting them loose with a knife.

14.—Taylor Frame.

Bars led to frames. Henry Taylor, in his Bee-Keeper's Manual, (first published in 1838,) 6th edition, London, 1860, *p.* 73, describes a frame like Fig. 14, and gives an illustration of which Fig. 15 is a copy. Describing his observing hive, he says:

" For the purpose of preventing the bees from attaching the combs to the glass, thin upright strips of wood, rather more than half an inch wide, are tacked under the centre of each bar, at both ends, extending from top to bottom inside of the hive. Or some might prefer to use frame-bars, like the one described and illustrated at page 58," as follows :

" It may be well here to allude to what some have thought to be an improvement in the construction of the bars, the object being to render the combs more accessi-

15.—Taylor Frame.

ble, and the usual cutting, to detach them from the sides of the hive, avoided. A reference to the accompanying engraving will exhibit a bar with a frame suspended beneath it, but so made as not to touch either the sides or bottom of the hive, and within which the combs are, or ought to be, wrought."

W. Augustus Munn, invented the " Bar-and-Frame Hive," and published a description of it in London, in 1844. He then used the " oblong bar-frames to take out of the back of the bee-box." He afterwards discarded the oblong frame, and in April, 1851, published a second edition of his pamphlet, describing his "improved hive" with his "triangular bar-frames, made to lift out of the top." Others have made their frames to slide in and out edgewise; in others, the frames partly lift and partly slide out edgewise, as in the "California hive."

PATENT HIVES.—FRAMES.

" The *Langstroth Hive*, like the *Huber* and *Munn* hives, is constructed on the movable-comb principle; but more properly combines the oblong-bar-frame, as originally used by Munn, with Bevan's bee-box, and other additional improvements, making it more simple and practical than either of its predecessors."—*J. S. Harbison in Bee Culture, p.* 149.

Mr. Langstroth says,* " I have before me a small pamphlet, published in London in 1851, describing the construction of the Bar-and-Frame Hive of W. A. Munn, Esq. The object of this invention is to elevate the frames one at a time *into a case with glass sides*, so that they may be examined without risk of annoyance from the bees.

" Great ingenuity is exhibited by the inventor of this *very costly* (?) and very complicated hive, who seems to imagine that smoke must be injurious, both to the bees and their brood."

Great as Mr. Munn's " ingenuity" may have been, it falls somewhat short of that exhibited by Mr. L., in the above quotation, by which it would appear that the hive of Mr. Munn was an *observation hive only*, whereas the *facts* are, that it was intended to combine all the desired advantages of a practical bee-hive for every day use. Mr. Munn says, *p.* 23, in speaking of other hives, " But however, I should not be doing justice to Mr. R. Golding, if I did not particularly mention his improved Grecian hive, by the use of which combs may be removed from the interior of the hive and inspected at pleasure." Again, same page, " *My object* has been to point out briefly to those anxious for the better, more extended,

*Honey Bee, 3d edition, 1860, page 209.—*Note.*

and economical mode of bee management, the difficulties to be provided against, and to recommend to their consideration the advantages offered in the bar-frame-hive."

A little history will make this matter more plain.

Mr. Munn's hive, with the " oblong-bar-frames," was described in a pamphlet, published in London in 1844. In April, 1851, a second edition was issued, in which the inventor refers to the " *oblong-bar-frames*, and introduces the triangular ones in their stead. On the fifth of October, 1852, Mr. Langstroth obtained a patent on " improvements in bee-hives," under which he is understood to claim " *all* movable frames in bee-hives !"

Mr. Otis, in one of the Langstroth circulars, says, " *This is the original movable* comb bee-hive" !!! and calls all those persons who claim the use of the movable comb frame, " PIRATES" !! [This latter clause is a dangerous weapon in the hands of the Langstroth men ; *it points the wrong way*.]

" Rev. L. L. Langstroth is the ORIGINAL INVENTOR of the movable comb frame" !! *C. B. Biglow, in Bee Journal, Sept.* 1861, *p.* 212.

In a small treatise, compiled by permission, from Langstroth on the Honey Bee, by Richard Colvin, the author says, *pp.* 36, 37, " Mr. Langstroth is the original inventor and *sole patentee* of movable frames in Bee Hives ! ! !"

We had hoped that these absurd pretentions of his agents had not fallen under Mr. Langstroth's eye, and had done him the justice to believe that, when brought to his notice, he would relieve himself from complicity in them by a disclaimer over his own signature ; but we regret to say, that further developments have seriously disturbed these favorable anticipations.

This matter is dwelt upon somewhat, because it is threateningly claimed by interested parties, as above cited, who perhaps(?) know no better, that the Langstroth patent secures to its holder the *sole* use of movable frames. For the benefit of such, and to aid the curious, we give herein a few of the facts, and subjoin an illustration of the Munn Hive.

> " With hasty judgment ne'er decide ;
> First hear what's said on t'other side."

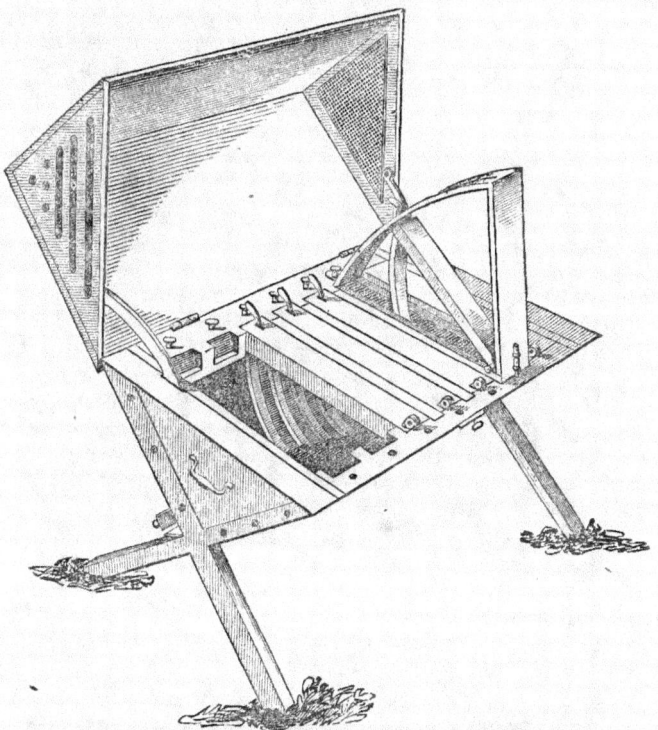

16.—The Munn Hive of 1851

Munn's Movable Frames.—In the preface to the pamphlet from which Mr. Langstroth quotes, as before cited, Mr. Munn says he has " very materially simplified the construction of the Bar-frame Hive, by forming the ' oblong-bar-frames' into ' triangular-frames,' and of the back of the Bee-box." Mr. Munn's mode of using his movable frames may be

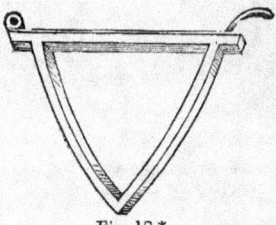

Fig. 18.*

seen by the following, from the same pamphlet. "The frames with their contents, may be lifted out into the 'observation frame' * * whenever it is wished to examine the bees, &c., as the half-inch spaces between the bee-frames, will allow of a sufficient distance to be preserved between the lateral surfaces of the perpendicular combs formed in the bee-frames, and thus permit them to lift out by each other with facility," *p.* 14. Again, "The whole interior of the hive is thus open to inspection at any moment, and a choice can be made of the combs containing the most honey, or the bee owner enabled to trace the devastations of the honey moth." *p.* 17. Still further, the hive should be so constructed as to allow of every part of the combs to be inspected at any moment, and capable of removal when requisite," *p.* 20.

We now give Mr. Langstroth's claim of movable frames:

"*Second.*—The use of the movable frames, A. A., fig. 4, or their equivalents, substantially as described; also their use in combination with the shallow chamber, with or without my arrangement for spare honey receptacles!"

Mr. Langstroth's frames, patented in 1852, are substantially the "oblong bar-frames" of Mr. Munn, described in 1844; and how his mode of using them compares with Mr. Munn's mode of using his triangular frames, described in April, 1851, as shown in the above quotations, the reader will be able to judge after consulting Mr. Langstroth's work on the *Honey Bee, pp.* 15, 148, 149, or examining one of his hives.

Munn's Divider.—On page 10 of his pa[m]
"One of the triangular bee-frames can be
can be used as a *divider* between any nu[m]
and thus form the box into two compartm[ents]

of 1851.

or diminish the space in the box, according to the size of the swarm, or the increasing wants of the bees for more room."

Mr. L's claim of the divider :

" *Third.*—A divider substantially as described, in combination with a movable cover, allowing the divider to be inserted from above between the ranges of comb." He says, " By means of a movable partition, my hive can be readily adjusted to the wants of either large or small colonies."—*Honey Bee, p.* 96.

Munn's space around the frames.—" The divider is made to fit close to the box at the two sides, by means of extra slips of wood, to prevent the bees crawling between the frames and outer box, as they can do around the bee-frames."—*p.* 14. " The bee-frames form, as it were, a smaller box within the triangular box, and are not in immediate contact with the external air, but *have a half-inch space nearly* all around them," *p.* 17.

Mr. L's claim :

" *First.*—The use of a shallow chamber, substantially as described, in combination with a perforated cover, for enlarging or diminishing at will the size and number of the spare honey receptacles."

What are the " essential and patented features" of Mr. L's invention ? Above we have given them as gleaned from his book, to which we have added his claims. On page 15 of his work, after describing and condemning the Huber, Munn, and all other hives, and having before him Mr. Munn's pamphlet of 1851, CONTAINING THE ABOVE ILLUSTRATION OF MUNN'S MOVABLE BAR-FRAME HIVE, Mr. L. says, " One thing, however, was still wanting. The *cutting* of the combs from their attachments to the sides of the hive [!!!] was attended with much loss of time, both to myself and the bees. This led me to INVENT A METHOD *by which the combs were attached to* MOVABLE FRAMES, so suspended in the hives as to touch neither top, bottom nor sides." !!!

In the *Bee Journal* for June, 1861, *p.* 142, Mr. L. says, " If Huber had only contrived a plan for suspending his frames, instead of folding them together like the leaves of a book, I believe that the cause of apiarian science would have been fifty years in advance of what it now is.

" Now if I had known that my hive was not so much better than Huber's as to deserve a patent, and if I had been base enough to attempt to palm upon the public substantially his invention as my own, can any man of common sense believe that I would have published to the world, just where and how I *stole* my pretended invention ?"

Of course not.

In the same communication, speaking of American movable comb bee-hives, he says : " *In my opinion* ALL OF THEM have appropriated to a greater or less extent, the essential and patented features of my invention." * * That he believes the courts will sustain this opinion, and that he should long since have sought their protection, but for his limited pecuniary resources, the state of his health, and the fact that *other parties* own the greater share of his patent ?

Having sold the territory, and got the money, he leaves the luckless purchasers of supposed " *rights*" to take care of themselves ! Consistency, indeed thou art a jewel !   But we will not quarrel with Mr. L. about that.   The concluding portion of his very interesting *Bee Journal* article, above cited, runs thus : " If any one can show that before my invention there existed any movable frame hive adapted to practical use, or any invention that used the essential and patented features of mine, I will try to be the first to acknowledge that although an *original inventor*, I was not the *first* inventor of such a hive."

Will he do it ?   We shall see.

Much more testimony of the same import might be given, but let this suffice.

I would not detract from the *just* claims of Mr. Langstroth ; neither am I willing that he should have credit for inventions not his own.   He has combined in his hive some of the most practical features of European inventions ; and he deserves to be, and has been well paid for the best *compilation* on the Honey Bee.   But justice to Bee-keeping, as well as to Mr. Harbison, compels me to add, that *he* is entitled to all praise for the best *original* work on that subject—notwithstanding its many errors—yet published in this country.

Although inventive skill nas not done all in this interesting field that bee-culturists desire, its achievements have, nevertheless, been of great value. The main features of success attained by the hives already noticed, are the well established convictions among intelligent bee-keepers—

1. That bees may be induced to build their combs with considerable regularity on frames put into their hives in a proper manner.

2. That they may be thus handled with safety to the apiarian and the bees.

3. That bees may be artificially swarmed, and losses by natural swarming prevented.

Their failures thus—

1. To provide a style of frames and method of using them that will not irritate the bees, nor injure them nor their combs.

2. To provide a convenient and reliable system of artificial swarming.

3. To provide safe and economical wintering.

Many and repeated experiments, by different persons, under a great variety of circumstances, and living in parts of the country remote from each other, have satisfied those who are acquainted with the facts, that in the hive and system now about to be presented, the three points last named have been reached, and that by their use much greater profits may be derived from bee culture than by any other hive and system.

THE MICHIGAN HIVES, invented by the author, consist of a Quadruple hive, a Double hive, and a Single hive.

19.—THE QUADRUPLE HIVE.

In the above cut, *A B C D* represent the ground plan of the four apartments of the quadruple hive, each of which, when full for wintering, should contain a good strong swarm of bees. The bottom board of the hive projects at the point *C*, or front, for an alighting board for the bees, the whole width of which is left open by removing the slide underneath the movable front as soon as warm weather approaches and blossoms are sufficiently

numerous to ensure safety from the depredations of robber bees. The movable front is secured in place by but-
tons on the side of the hive, and held up by the cleat on
its front, resting on cleats on the permanent sides of the
hive. When the hive is all together, this cleat, or stay,
runs all the way around the hive, and also serves as a
support to the top, or surplus honey chamber. The
whole hive rests on a cross, made equal to the width
of the hive. A centre point projects from this, fitted
to a hole made half way through the bottom, on which
the whole *turns*. At the point d in the frames, cylinders
of tin or other suitable material, five-eighths of an inch in
diameter and of the same depth, are placed for the purpose
of securing an opening through the combs at this point,
when they shall have been constructed all through the hive.

The manner of operating with this hive depends somewhat on what we wish to do with it.

If we desire only to double the number of our stocks in
a single season, and hence secure the largest amount of
surplus honey, as soon as bees have begun their labors in
the spring we transfer bees and brood combs to other
hives, leaving but two stocks in the premises, and occu-
pying opposite apartments, as follows :—

Early in the morning we close in all four of the swarms,
and gently remove the old hive a little off its stand, and
put a clean one in its place in exactly the same position
the old one occupied. We blow a whiff or two of smoke
of burning cotton rags, wood, or tobacco, in among the
bees to alarm them. They at once fill themselves with
honey, so that when we open the hive, if we are deliberate
and careful in our movements, they will not sting us.
We now transfer the swarm in A, by means of the
movable frames, into the department A of the new hive ;
placing the frames always in the same relative position

in respect to the hive and each other that they occupied in the old hive, and so adjusting the inner passage ways that the bees of *A* may pass through the empty part, *B*, in going to and from the fields, while they also use the larger and more direct outlet at *a*. So, also, we transfer the stock from the hive *C* to the corresponding apartment of the new hive, leaving this swarm to work out of *C* at *a* and through the empty apartment *D*, in like manner ; the two swarms being thus entirely separate from each other. The movable fronts in the tenantless apartments *B* and *D* are now left out, and the top cover is put on to protect the hive from the weather. We now provide a *new stand* and hive for the remaining two swarms, which are to be transferred in the same manner, placing them in the new location. Then we *renovate the old hive*, cleaning it with water boiling hot, so as to remove any gum which the bees may have placed upon the gauze wire curtains in the partition walls, and destroy every vestige of the moth that may be lurking in any crevice of the hive. Any needed repairs should now be attended to. Our bees are thus put into summer quarters and are ready for

ARTIFICIAL SWARMING.

This is done as soon as the drones—the male bees—make their appearance, and in the following manner, viz. : We open the hive *A*, and selecting a comb, containing

20.—Brood Comb.

brood in all stages of development, from the egg to the capped larva, as shown in Fig. 20, we transfer it, with its adhering bees (being careful not to get the queen), to the empty part *B*, putting in its place an empty comb frame, and shutting off the communication between the apartments *A* and *B* at the point *b*. This card of comb we *lean over* a trifle, so that it shall remain firmly in position, putting in place the movable front and slide so as to leave only the small opening, like that shown at *a*. We thus detach a small portion of the bees from the parent stock, *A*, for the purpose of rearing queens, which they will immediately set themselves about doing from the brood furnished them. Such of these bees as have been in the habit of using the hitherto empty apartment for a passage way, will remain, to be joined from day to day by numbers from the parent hive that have also used that apartment as a passage way ; while those we have transferred, that have been accustomed to use the direct outlet from *A*, will of course return thither. We thus secure only a comparatively small portion of the bees for the purpose of queen rearing, while the labors of the parent stock go on undisturbed. Indeed, the old hive will breed *all the faster*, for the reason that the vacant space

21 —Queen Cell Inserted.

will be rapidly filled with new worker comb, in which the queen will immediately deposit eggs ; while in the infant colony no combs are constructed at all, because the bees have enough already. The tenth day after starting the queen cluster, if no royal cells had been begun when the comb was transferred, we open the hive *B*, and with a sharp penknife *cut out all but one of the queen cells*, using these immediately in starting other queen clusters — attaching one of them to a card of comb and bees (fig. 21), taken from the hive *C*, and transferred to *D*, in the manner before described ; being careful always to *cut off the communication between the new queen cluster and the old swarm,* or the queen will not be allowed to mature, and the bees will return to the parent stock.

In transferring queen cells, it is of no consequence how we place them in the combs, so that we do not injure them. I have placed them in every imaginable position with equally favorable results. Great care must be taken not to press them with the fingers, nor let them lie in the sun, or exposed to the chill of morning or evening, for fear of destroying the royal occupants. The tip of the cell should not touch the comb, as, if it does, the bees may stick it fast at that point, and thus prevent the hatching of the queen. The inexperienced bee-keeper had better transfer only one of the queen cells at a time, returning the frame from which it is taken to its place in the hive till the royal cell is properly adjusted in its new location, in order to prevent injury to the young larva. Each new colony should receive *only one* queen cell, because it is found that a queen emerging in a small colony, with no rivals in prospect, will make her excursion to meet the drones several days sooner than one emerging in a populous colony, or having rival queens in prospect to be disposed of.

The following letter from the author, was published in the *Bee Journal* for September, 1861, *p.* 212 ; and the editor, in his remarks on the same, clearly endorses the views here set forth :

A great diversity of opinion exists as to the time when the first excursion of a young queen in quest of drones for impregnation may be looked for. The June number of the Journal (page 130.) states the time at from the fifth to the twelfth day after issuing from the cell. I think this is a mistake ; at least it has not been true with me.

I have practised *artificial swarming* exclusively, and made a record of the facts. The queen may be *confidently* expected to issue from the hive, between noon and half-past two o'clock P. M., on the second day after emerging from the cell—frequently on the first, —and if drones are abundant, she usually meets them after one or two flights.

A practised eye will readily recognize the marks of impregnation with which she returns when successful ; and in from two to ten days thereafter she will generally be found depositing eggs in the cells. One queen which issued from the cell on the 4th of July, took wing on the 5th, and had deposited quite a quantity of eggs on the 7th. Out of six which issued on the 26th ult., three became fertile on the 29th, two on the 30th, and one on the 1st inst. These are instances of the *earliest* fertility, however, I have ever known. It is accomplished in the following way, viz : *by permitting only one queen cell to remain in the hive*. In rearing queens, I always use *small clusters* only. If more than one queen be allowed to mature, and the swarm be large, the bees are apt to *cluster around and imprison the queens :* besides this, the queen will destroy all surplus cells before leaving ; which, it is imagined, delays her impregnation. I have known the bees to thus imprison a queen for *ten days !* By allowing *only one* royal cell to remain in the hive after the tenth day, no such result will ensue. The only difficulty in thus rearing queens by small clusters, in warm weather, is the greater liability of the bees to take flight with the queen when she seeks the drones, and then leave for the woods. This source of vexation and anxiety, is avoided by taking the precaution of having some larvæ or capped brood in the cells at this time. The bees will not then desert their nurselings, and the queen will return—except an occasional one. A few will be lost by accidents, such as being destroyed by birds, &c., to which risk *all* queens are *once* exposed.

Grand Rapids, Mich., Aug. 4, 1861.

EDITOR'S REMARKS.—-There is greater diversity in this important matter than is commonly supposed ; and observers may differ widely in their statements and inferences, while each narrates the

facts correctly. Circumstances exert a controlling influence and
materially affect the result. Thus queens reared in small nuclei,
such as our correspondent uses, will certainly issue earlier and usually become fertile sooner, than such as are reared in larger colonies: and the seasonable removal of all surplus royal cells, will
efficiently contribute to bring about the desired consummation.
On the other hand, the young queen of a populous colony, whose
hive was *full of comb*, well supplied with brood and honey, has been
known not to be impregnated, though drones abounded, till more
than three weeks after she left her cell. The truth seems to be,
that there is no definite term—circumstances governing in every
case.

After we have thus adjusted the new colonies, we let
them remain for from six to ten days, when, if drones are
abundant, and we have safely transferred the cells, we
shall probably find our queens have become fertile, and
have commenced the work of depositing eggs. We now
catch the queen—she will not sting—between the thumb
and finger, and with a pair of scissors, *clip one wing*, so
that she cannot fly. This is to guard against losing a
swarm at a future time, should we neglect to swarm the
bees, or give them work to do. We also now *cage the
queen* for about three days, by placing her in a case of
gauze wire cloth a little larger than a thimble, and suspended in the hive, or laid upon the top of the frames
through one of the holes in the honey board, while we
are swarming the bees. After thus securing the queen,
and filling up the hives with empty comb frames, we turn
the whole one-fourth the way round, thus causing the
parent and infant colonies to exchange places, throwing
out of the parent stocks swarms of worker bees, into the
infant colonies. The hive should not be turned between
the hatching of the young queens, and their fertilization,
because *bees belonging to swarms of fertile and unfertile queens*
will not fraternize, but will quarrel. It might be turned

with safety a day or two before the hatching of the
queens, but it is more difficult to find the queen among
the greater number of bees ; hence, soon after her fertilization is the best time. It may *sometimes* happen, when
this operation is performed at a time when the honey harvest has received a check from a storm or otherwise, that
the bees, thus empty of honey, and consequently more
quarrelsome, suddenly thrown into the presence of a
strange queen, are inclined to sting her. It is to prevent
this, that she is caged for the space of three days, after
which she may safely be liberated. The bees cannot
harm her through gauze wire cloth not coarser than four-
teen meshes to the inch. The swarm will suffer no particular detriment by her confinement, since comb building
will go on as if she were at liberty. But this is only a
precaution to beginners, the experienced apiarian will
always know when to cage the queen ; since in the midst
of the swarming season, when the honey blossoms are
yielding in profusion, little or no precaution is needed to
protect either the queen or the operator.

Where great rapidity of multiplication of swarms is the
object, *one stock only* is left in the quadruple hive in spring,
leaving out, in this case, the movable fronts of all the un-
occupied apartments, and opening *all* the passage ways
through the inner walls. We now transfer a card of
comb, bees, and brood from *A* to *B*, proceeding as before
described. The tenth day thereafter, *from some other hive*,
take two more cards of comb and bees, for *C* and *D*,
giving to each of these a queen cell, taken from *B*, and
always capped over. We should use no other, as the bees
will be likely to destroy them. When our young queens
have matured, we turn the hive half the way round, letting it thus remain from eighteen to twenty days, or un-

til the hive *C* is nearly filled with combs. We now turn
the hive one-fourth, and swarming is done, having from
one good swarm thrown off three new ones by the time
natural swarming has commenced !

There are other methods of swarming the bees by revolving the hive, which will suggest themselves to the
apiarist, and by which he may be able to multiply them
to any extent. Let no one misunderstand, however, and
expect from fifty to one hundred pounds of surplus honey
from each of them ; for it is an extraordinary year indeed
for honey, that will afford a supply sufficient for wintering, where *three* swarms are taken from one. Bees must
have stores to live on through the winter, like everything
else in our climate, and it should be remembered that *they
are first* entitled to their stores, and *we* should be content
with the *surplus* for our care and attention.

Sometimes the combs become so filled with bee-bread
and honey, that there is not sufficient room for breeding
faster than the bees perish,—hence the hive proves un-
profitable. An exchange of combs, giving empty for full
ones, will rapidly augment their numbers. Every good
piece of worker comb should be saved for this purpose—
they can be readily attached to the top of the frame, by a
little melted bees wax applied with a feather, or the edge
of the comb may be dipped in a little melted bees wax, and
then placed quickly on the frame. If plenty of such
empty *worker* combs be furnished them early in the season,
at a time when comb building is conducted the most
slowly, they will be immediately filled with eggs, insur-
ing the multiplication of the bees with the greatest rapidity ; so that when the honey harvest comes, a supply
of laborers will be on hand to collect it. Those who
know how rapidly bees breed under favorable circumstan-

ces at this season of the year, and in how incredibly short a space of time their abundant stores are collected immediately thereafter, will appreciate the advantages thus secured. If empty combs are not at hand, give empty frames, letting them alternate with full ones, so as to secure true, even combs. In my process of artificial swarming, bees build *all their combs true,* for the reason that a guide is furnished them, which is a frame filled with even comb. On this they cluster, building the first new comb parallel with the one furnished them ; this becomes the guide for the next, and so on till all the frames are filled. *Worker* combs are secured in consequence of having a young queen, drone comb being seldom built in any hive during the first year of the queen's existence.

DOUBLE HIVES.

The Michigan Double Hive consists of two apartments, with entrances at the middle of each end, and gauze wire curtain and bee passage between the two apartments. In swarming artificially, this hive is to be turned one-half round, making the two apartments change places precisely. Those who keep only a few swarms may prefer the double to the quadruple hive, as it is more simple in management ; but in wintering, the quadruple has decidedly the advantage over the double hive.

If the directions herein given are followed, the hive will work, and beekeeping prove satisfactory ; but let no one flatter himself that because he has a patent hive his bees will take care of themselves and the improvident bee-keeper besides, and endure our long winters without honey. Bees may die of starvation in any hive, however well constructed.

SINGLE HIVES.

Fig. 22 is a side view of the hive, showing the manner of adjusting the frames. The single hive is made exactly

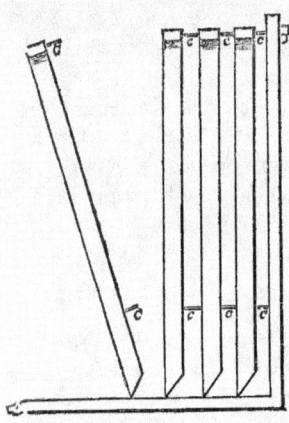

Fig. 22.

like, and of the same interior dimensions of, one-quarter of a quadruple one ; that is, twelve by twelve inches, and seventeen inches high. Eight frames fill the hive. Those who prefer inclined bottom boards, can rest the frames on wedge shaped cleats secured on the inside, and near the bottom of the side walls of the hive. This will allow the bottom to be hung by hooks, and swung up close to the lower edge of the movable front, dispensing with the slide. This style of hive, with my improved frames, possesses many advantages over any other single hive. It is simple, easily made, and cheap ; readily cleaned of all filth, by swinging the bottom board down, without in the least disturbing the bees, affords not the slightest point, inside of the hive, inaccessible to them, for the moth to deposit her eggs,* and combines the most practical form of the movable comb frame, and manner of using it, in a plain box hive.

* "There being *no* such thing as a *moth-proof hive* in *existence*, nor any prospect of such a discovery ever being *made*, we are compelled to be content with that which makes the nearest approach to it, viz., one that gives the bee-keeper easy access to the worms."—*Bee Culture*, *p.* 115.

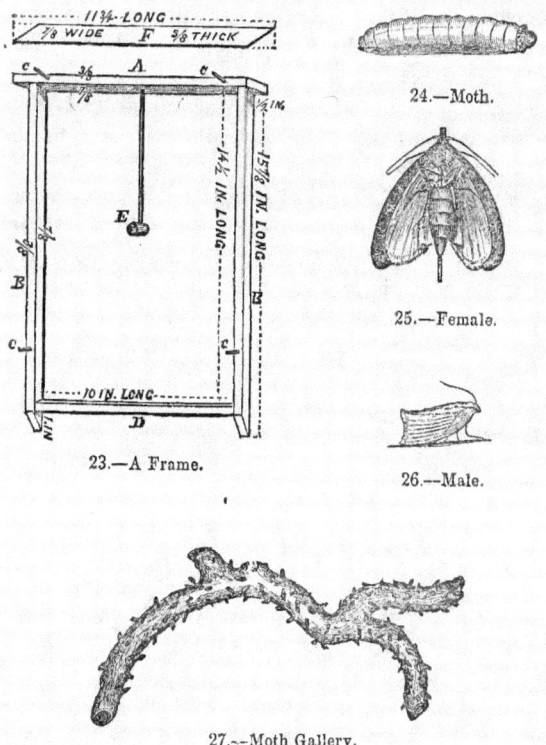

24.—Moth.

25.—Female.

23.—A Frame.

26.—Male.

27.—Moth Gallery.

Fig. 23 is a frame. *F*, plan view of top bar. *E*, cylinder for winter passage. The figures give the dimensions.

Fig. 24 is the moth worm ; Figs. 25 and 26 the winged moth, or miller ; and Fig. 27, section of moth gallery. The female is somewhat larger than the male. She enters the hive at night and deposits her eggs, preferring to leave them on the brood combs. As soon as hatched, the worm encloses itself for protection in a silken case, which, extended, becomes its gallery, or course, through the comb along its central wall.

The nearest approach to a "moth-proof" hive, is one so constructed that the miller can find no crevice in its interior, to which the bees do not have access, in which to deposit her eggs. Where two pieces of wood come together is the place sought by the moth, thrusting in her ovipositor and leaving the egg to hatch and begin its gallery beyond the reach of the bees. In passing from this point through the hive and combs, the worm continues to spin its silken protection, which is proof against all assaults of the honey bee. Once safely within the comb, the moth, protected by its gallery, passes along the wall at the base of the cells, sticking fast in its silken toils the unhatched bees. I have seen thousands of them just ready to emerge, vainly struggling to free their extremities from the grasp of the destroyer; these soon perish. The only remedy is, to cut away the worm gallery and remove the dead and dying brood. The moth does not directly destroy the brood, but only feeds on the wax of the cells and the food deposited therein for the development of the young bees.

A strong, vigorous stock, having a fertile queen, will not allow the moth thus to get possession of the hive; and if, while destitute of a queen, a foothold is gained, the bees, on the maturity of the young queen, will cut away the comb possessed by the worms, letting it fall to the bottom. They will then carry from the hive by piece-meal such portions of it as they can separate from the mass, plastering over the remainder, if any, with their propolis. Swarms not sufficiently populous to cover all their combs, and especially queenless ones, are most exposed to the moth ; and old black comb is more liable to be destoryed than new.

The removal of the moth gallery by the bees subjects them to great labor and much loss of time, which the use of movable frames will entirely obviate, as they will enable the bee-keeper to inspect the combs at any time, and remove the worms and any portion of the comb occupied by their gallery ; or give any other relief that the swarm may need. Hence writers on bees are agreed as to the necessity of using movable frames, as a means to successful bee-keeping. The only open question in the matter is, the style of frames and the manner of using them. There should be as little contact of surfaces inside the hive as possible : hence it is obvious that the frame which has the least bearing in the hive, and makes the least crevice inaccessible to the bees, is the best for this purpose. Some inventors of hives are aware of this ; and one of them says that in his hive, such a place is found only " where the frames hang in the rabbets."

My hive is so constructed that *no* crevice is found in its interior in which the miller may deposit her eggs beyond the reach of the bees, not even where the frames touch the hive.

There is no " moth-proof" hive, and cannot be ; for the reason that the miller will go anywhere that a bee can. Many ingenious devices have been invented for excluding them. For example : A " pedal" is fitted in the entrance intended to be operated by the weight of the bee, but so nicely adjusted that the lighter body of the miller will not open it. Beautiful in theory, but worthless in practice ; for the reasons, that in hot weather the bees will lie in and about the entrance night and day, keeping it open all the time, and soon cement the "pedal" immovably fast with their propolis.

SURPLUS HONEY BOXES.

When swarming is done, the honey boxes should be placed upon the hive. These should have a bottom with holes to correspond with those in the cover, or honey board, so as to be readily removed. Honey boxes should never be put on the honey board WITHOUT BOTTOMS, nor on the frames without a honey board. If they are, it is difficult to remove them without injury to the combs. If glass sides and ends are made to the boxes, so that the honey can be seen, it will sell for enough more in market to pay the extra expense. Two boxes should be placed upon each hive, or swarm, six inches wide by five deep and twelve inches long, as represented in Fig. 19, p. 36.

WINTERING.

To prevent loss of bees by starvation in winter, with plenty of honey on hand, is the object, in part, for which the quadruple hive and the WINTER PASSAGES are constructed. In a single hive, without a passage through the combs, it frequently happens, that during cold, freezing weather, long continued, the combs outside the cluster of the bees, become covered with frost, the congelation of their breath, which they are totally unable to remove—and they will not go over it—and SO PERISH OF HUNGER AND FALL TO THE BOTTOM BOARD WITH PLENTY OF STORES ALL AROUND THEM ; and while the bees between the ranges of comb are in prosperous condition, in consequence of the greater degree of animal heat nearer the centre of the cluster. In low, broad hives, even with winter passages provided, whole swarms often perish in like manner, having eaten their way to the honey board.*

With the thermometer at or near zero, a large swarm of bees will cluster in a circle of eight or ten inches. Hence it becomes evident, that in a hive of proper proportions, with suitable winter passages, the bees will pass through them to the interior of the hive—the swarm thus expanding and contracting as the cold diminishes or increases. For similar reasons bees winter more safely in the four-colony hive than in either double or single ones, where both are alike exposed to the winds and frosts of winter.

Ventilation.—A ventilating passage should be secured for a supply of fresh air, even during the coldest weather. This may be done by making a small bee passage through the movable front near its top, like that at its bottom, and leaving both open. A *better* way, however, to ventilate a swarm of bees in my hive is, to place in the fall, clean, gauze-wire cloth over the holes in the honey board, and fill the top chamber with fine straw, chaff, shavings, or other dry, porous material. This will allow sufficient air, and at the same time absorb the moisture

* " The Langstroth hive had also been introduced into a number of apiaries, ours among others. From the glowing accounts which I had heard of it while in California, I expected to find the desideratum long sought for by apiarists, and as a result of its introduction into our apiaries, that they would be in a highly flourishing condition, particularly that portion of the stock contained in the new style of hive. In this I was doomed to disappointment, as most of the bees that had been put into them had died of starvation, they having eaten all the stores from the bottom to the top of the hive, in the centre of a diameter equal to the size of the cluster, leaving an abundance of stores still within the hive, but owing to the severe cold, the bees were unable to reach them."--*J. S. Harbison, in Bee Culture, p.* 31.

contained in the breath of the bees, keeping them dry and sweet, and preventing a current of cold air through the hive, which is fatal.

The slide should occasionally be removed, and all dead bees and dirt drawn out from the bottom. A convenient scraper for doing this may be made of a 3-16th inch brazier's rod sixteen inches long, and the end turned about one inch and flattened.

Bees thus cared for, in my hives, and placed where the sun shall not shine upon and disturb them in the middle of the warmest days of winter, WILL NOT PERISH WHILE THERE IS HONEY IN THE HIVE

By those who do not keep bees in such numbers as to render such a course impracticable, something may, perhaps, be gained by carrying the hives in early winter into a *dry, dark, quiet cellar.* The bees will thus remain more quiet, and consume less honey, than otherwise ; but this costs time, and is attended with care and trouble ; and, besides, most cellars are so damp as to render the destruction of the bees certain : so that it is not probable that this method could be adopted to any considerable extent.

TALL HIVES.

Intelligent bee-keepers are generally agreed that tall hives are better to winter bees in than low ones. Even Mr. Langstroth, whose hive is a low one, says : "A hive *toll* in proportion to its other dimensions, has some obvious advantages ; for, as bees are disposed to carry their stores as far as possible from the entrance, they will fill its upper part with honey, using the lower part mainly for brood, thus escaping the danger of being caught, in cold weather, among empty ranges of comb, while they still have honey unconsumed."—*Honey Bee, pp.* 329, 330. Mr. J. S Harbison to the same effect: "Many eminent apiarists bear testimony to the superiority of deep hives over those that are low and of large diameter."— *Bee Culture, p.* 132.

Mr. Langstroth's *frames* compel him to forego the "*obvious advantages*" of a tall hive. He says (Honey Bee, p. 330) : "It would be impossible to use frames in it to advantage"—true in regard to his frames and his mode of using them—and in a foot note he gives the following very good reason : "The *deeper* the frames the more difficult it is to make them hang *true* on the rabbets, and the greater the difficulty of handling them without crushing the bees, or breaking the combs."

In the Michigan quadruple hive are combined the advantages of both height and depth—the frames being so constructed and operated as to admit any desired height of hive, and the main entrance of each apartment being at the corner most remote from the centre, around which the bees, each colony in its own apartment, cluster in winter. The quadruple hive gives each colony the benefit to be derived from a single one equal in length to the diagonal of the quadruple, in addition to the advantages of nearly four times the lateral space to be had in an ordinary single hive. Hence the quadruple hive is adapted to fully gratify the "disposition of the bees to carry their stores as far as possible from the entrance," a feature not attainable in any other hive.*

---

* Those who have bees in low hives, will find they will winter better if the hive be set on its end late in the fall, keeping the combs in a perpendicular position. Before doing this, the frames and honey board should be made fast, so as to be kept in place.

GREATER ANIMAL HEAT.

In the quadruple hive, only one-half of the wall surface of each apartment is exposed to the weather ; and, in winter, the four swarms cluster about its centre, thereby producing in the hive, for the benefit of each swarm, four times the amount of animal heat produced in a single hive.

WINTER PASSAGES.

Many colonies of bees are lost in winter, from want of *winter passages* through their combs. Seeing this necessity, the writer contrived his present mode of making such passages, and securing them against being filled up by the bees, by cylinders made of tin or other material, and painted on the inside, and suspended in the empty frames, or placed in the combs.

Mr. Langstroth, in his book on the Honey Bee, third edition, 1860, *p* 337, recommends cutting a hole through each comb late in the fall ; and in a foot note, gives Mr. Wm. W. Cary's method of making such a passage, describes his instrument for doing it, says an application for a patent on " this device" was pending, and that, " if the patent issues, the right to use it, will be free to all owning the right to use the movable-comb hive."

It may be remarked, here, that he says nothing about securing the passage against being filled up by the bees. Yet, in the *Bee Journal* for June, 1861, *p*. 136, Mr. L. says,

" *Some years ago* Mr. W. W. Cary, of Coleraine, Mass., after cutting winter passages in the combs, put in them a coiled shaving, to prevent the bees filling them up. I contrived a mode of *suspending* this shaving in an empty frame."

It is a *singular coincidence!* that between the two writings of Mr. L., my hive, with its winter passages, had fallen in his way, and that at the time of his latter writing, my patent, embracing said winter passages, had already been ordered to issue! Mr. Langstroth's *"some years ago"* were included between 1860, and June, 1861 !!!

"DYSENTERY."

Once during winter, it is desirable, and in long winters quite necessary, that bees should be allowed to fly, to discharge their fæces, or they are apt to be attacked with what is improperly called "dysentery." This arises from the inability of bees, after long confinement, to retain their fæces, consequently on the approach of a mild day, at such times, even when the weather is too inclement for them to safely fly, many will venture out for this purpose, and drop down upon the snow, while some evacuate about the entrance and *in the hive.* When the latter takes place to any considerable extent, the whole swarm is aroused to great activity; and, if the weather continues cold, perish. A swarm in this condition should be given air, and carefully shaded. As soon as the thermometer marks 45° in the shade, place them in the sun, and open the hive to let them fly. I have saved swarms in mid-winter by allowing them to fly in a room, setting the hive by the window, and returning them with a ladle or spoon. But a room is quickly so soiled by the bees that it is not fit for anything else. After the bees have thus relieved themselves, the disease disappears.

FEEDING.

Bees should not be fed with liquid sweets in winter, when it can be safely avoided. Swarms, light in the fall, should be united, or their insufficient stores replenished, by cards of comb well filled with honey from other hives ; or, a box of honey *with its bottom* and the honey board removed, so that the bees shall have ready access to it, may be placed on the frames. They may *starve* with it above the honey board. Bees may be fed *in the fall* to some advantage, when the swarm happens to have been started late, or removed from the woods ; candy, or anything of that nature, may be placed immediately above the frames and *accessible* to the bees *in cold weather ;* or liquid sweets may be given them, but I have always found *sealed honey the best and cheapest bee food.*

They do *not* need water, as some suppose, unless we want to encourage breeding, which is not advisable in winter, as it causes them to use more honey. But in spring, for breeding purposes, a considerable quantity both of honey and water is needed. Even after blossoms appear, if the weather continues for several days too cold and stormy for them to fly, they will often perish if not fed. A sponge kept saturated with sweetened water, placed on the wire curtains covering the holes in the honey boards, will save them, and in any case do no harm.

ROBBERS.

Should robbers be enticed thereby, or at any time, from any cause, contract the entrance, and if they still persist, close it up, so that but a single bee can crawl through at a time. This will give the defenders of the place the best of the fight, and they will soon rid the premises of their assailants.

How to take them.—It sometimes happens that a powerful swarm from the neighboring forest attacks a weak swarm and nearly ruins it before discovered. In such

case, close up the hive entirely ; place by its side a hive
having within it a card of honey, or comb filled with
sweetened water ; let the bees come and go a few times,
and they will *fairly swarm* about you, encouraged by their
success. When in the midst of their labors, place a tube
in the hive, fitted to the bee entrance, and long enough
to reach about half way through the hive, with its inner
end elevated a little from the bottom. Now open one
side of the hive, so as to let in the light—it should have
one side of glass covered with a shutter—until the bees
have filled and want to go home, when they will fly for
the light, and find themselves trapped. Having provided
a hole in the top of this hive, which can be opened and
closed by means of a gate, place a hive on the top of
this, containing a piece of brood comb freshly taken from
a hive. Open the gate, close the shutter and entrance
below, and open the shutter of the upper hive, until the
bees, thus caged, have ascended into the upper hive ;
then close the gate and the shutter of the upper, and open
the shutter and entrance to the lower hive, letting in
another band of robbers ; and thus continue till you
have caged the whole gang. The hives are now to be removed ; the top one to a permanent place in the apiary,
where it is to stand till about an hour before sun down
of the fourth day. It is then to be opened, and the bees
given their liberty, when it will be found, that several
queens have been started, and your robbers have con-
cluded to stay with you altogether, and you can count
one more swarm in the apiary. If small tin valves are
placed upon the inner end of the tube, to be operated by
the bees themselves, they are more easily caged, and
without the gate and shutter. I discovered this device—
the valves—about two years ago, and thought it new, but

found that Mr. R. B. Merritt, of Battle Creek, Mich., was ahead of me, although inventing it for a different purpose. He did not patent it, however, and I believe it is now public property, any one having a right use it. The idea of thus starting a swarm of bees, I believe to be original with myself. I first practiced it about six years ago, by catching a few bees from the blossoms in the fields for the purpose of experimenting, not thinking, at the time, to what use the principle thus demonstrated might be applied.

TRANSFERRING FROM BOX HIVES TO MOVABLE FRAMES.

Having provided a box, called the " driving box," so made that its mouth will exactly fit the open end of the hive from which the bees are to be driven, blow a few whiffs of smoke from burning wood, cotton rags, or tobacco, in among the bees—not too much so as to sicken and cause them to fall down from among the combs upon the bottom board, but only enough to alarm and induce them to fill themselves with honey. Let the hive stand five minutes, to give the bees which may be out gathering honey, time to return ; then blow in a little more smoke, driving them all up among the combs. Now remove the hive two or three rods from the apiary, under a convenient shade, if such is at hand, and carefully turn it over on a clean spot, (but never upon loose *dirt or cultivated ground*,) with its bottom or open end upwards. As quickly as possible cut a small piece of brood comb, or comb containing eggs and young bees, from near the centre of the hive, and suspend it by a nail, or otherwise, in a box, for a temporary hive, placed upon the stand from which the box hive was removed, to catch up such bees

as may be out seeking their old hive.   These will cluster
upon the comb thus furnished them, till they are wanted,
and be prevented from entering other contiguous hives.
Place the driving box upon the box hive, so that they
will exactly fit each other, mouth to mouth, tacking the
two together with a couple of nails, and with rags close
every crevice, so that not a bee shall escape.   Now
lightly rap on the top—now bottom—of the hive, gradually moving up, from fifteen to thirty minutes.   By this
time nearly all the bees will have ascended into the top
box, which will be known by the humming noise within,
on applying the ear to the side of the hive.   A window of
glass, or wire cloth, provided with a shutter, may be inserted in the side of the driving box, through which to
see the bees.   If the driving box have sticks nailed across
its interior, for the bees to cluster upon, all the better.
When the bees have nearly all ascended into the top box,
it is to be removed, and a cloth, or wire curtain, open
enough to give plenty of fresh air, is to be placed over
its mouth, to prevent the escape of the bees, letting it
stand in a cool, shady place while transferring the comb.

After removing the side of the box running nearest

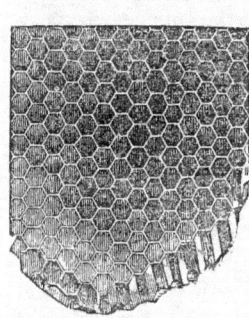

parallel with the comb, as carefully as possible, cut out one card,
placing it upon a common tea
salver, and with a frame lying
upon it in such a manner that the
honey and breeding cells shall remain in the same relative position in the frame that they occupied in the hive, as in Fig. 28,
cut the comb a trifle larger than
to fit, so as the better to fasten it

28 --Comb fitted for frame.

in place, using only the truest, evenest combs, and dis
carding all drone cells and combs of much greater thick
ness than one inch.

Old black combs that are true and even, are just as
good as new, white ones. Having thus fitted the comb
to the frame, tack on each side a strip of wood, previ-
ously prepared, of about $\frac{1}{4}$ by $\frac{3}{8}$ of an inch material, and
long enough to reach from the bottom to the top of the
frame, to hold the comb in its place for three or four days,
or until the bees shall permanently attach it to the frame.
If a little melted bees wax is at hand, the combs may be
partially attached with this by means of a feather, but
except during the midst of the honey harvest, the fumes
from burning wax or combs should be avoided on account
of greater liability to entice robbers.

When thus prepared, the frame, with its comb and
brood, is carefully placed in the hive, and its entrance
closed up to keep out inquisitive stranger bees. Thus,
one by one, are the combs removed, without honey, except what little there may be near the corners of the
frames, the surplus being immediately taken to the house
out of the reach of the bees. Having thus placed all the
combs in position in the new hive, the small box is
brought from the old stand, and the bees it contains shook
out at the entrance of their new domicil, having first removed the slide, so as to give them room to enter. A
little smoke may be needed before removing it, as the
bees in it have probably already begun the work of queen
rearing, and may be loth to leave the comb furnished them,
which should now be removed. If we intend to *practice
artificial swarming*, and prevent our bees from absconding
to the woods, we have now to *find the queen*. This is
easily accomplished after having been once done, and

the *inexperienced* will have but little trouble to find her
by turning up the driving box on its side, and gently
dipping the bees with a spoon, or ladle, from *that* to the
entrance of another box, after the manner of hiving a
natural swarm, looking over each spoonful carefully for
the queen. A few of the first spoonfuls may turn back
toward the greater noise in the driving box ; but perse-
verance and a little sprinkling of water, will soon get
them started the other way ; and if a bee is seen a trifle
larger around the body and *nearly twice the length* of a
worker, and considerably less bulky than a drone, you
may be sure you have the queen. If she is not found the
first time, exchange boxes and search again. When
found, *clip one of her wings,* so that she shall be unable to
fly. This will not impair her usefulness in any degree,
for her wings are now of no use whatever, except to lead
off a swarm to the woods. But, it should be remembered,
that it is safe to clip the wings of FERTILE QUEENS ONLY.
ALL queens become fertile, if at all, within the first twenty-one days of their existence ; otherwise they are drone
layers. For this and other reasons, previously noticed,
it is not safe to transfer a swarm, with their combs, from
box hives at any time *between the issue* of *a first and any
after swarm* of the current year ; nor to "drive" a swarm
at such times, for similar reasons ("Why?"), because,
either no young queens are yet hatched, and we run
the risk of destroying them in their cells, or they are
unfertile, since they do not fly to meet the drones until
after leading off their swarms, and if we drive out the only
queen in the parent colony while there are no eggs in the
combs from which the bees can rear another, its final destruction becomes inevitable. The reasons are here given,
that the inexperienced may not fall into an error, often

committed, and not immediately remedied as it should be, by supplying a fertile queen from another hive, leaving the stock from which she is taken to rear a new one. Another reason why it is unsafe to transfer a swarm of bees while they have yet an unfertile queen is, that a disturbance of the swarm at this time is apt to cause the bees to accompany the young queen upon her bridal trip, sometimes returning, but often absconding with her to the forests. This result may be prevented by timely giving such swarms brood comb, containing eggs and larvæ, from other hives. It is also an effectual remedy against *any* loss of swarms by flight after hiving ; for *bees will not desert their young*. In four to six days after transferring, the combs are to be righted, and the temporary slats removed, and any needed correction of position attended to.

"DRIVING," OR FORCED SWARMING.

This operation is performed precisely as if you intended to transfer the combs, with this difference only, namely : when you have separated the bees from the combs contained in the old hive, cover the latter with a gauze-wire cloth, or other suitable material, while the queen is being hunted up. When found, she is placed with one-fourth of the bees in a new hive upon the old stand, while three-fourths of the bees are taken with the old hive, and placed in a new location. If this operation be performed ten days before others are going to be driven, queen cells may be taken from this for supplying other hives. But to swarm bees in this manner, requires considerable skill, and a good degree of knowledge of bees ;—and even then is liable to fail, from the fact that we work in a great measure in the dark, and can never know, as we ought, whether a new *fertile* queen is provided or not, until it is quite too late to remedy the evil, if one be wanting. In thus transferring bees and combs from box hives to movable frames, we have purposely left them unsupplied with honey to any extent. When transferring is performed in the midst of the honey season, none is needed ; but if, as is often the case, we change bees to new hives late in the season, they may be unable to collect a supply of stores for the winter ; especially will this be the case if considerable empty space is left to be supplied with new combs. In such cases, a box four or five inches deep should be provided, without top or bottom, made twelve by twelve inches inside, so as to fit the hive. After the combs in the hive are righted, this should be placed on the top of the hive, making a chamber above the frames 12 by 12 inches and 4 or 5 deep, into which, properly spaced, a supply of sealed honey should be placed, and the honey board put on. Twenty-five pounds of honey so provided, will safely winter a large swarm any where that bees can be wintered, even though no other stores are in the hive. When apple trees are in full bloom, remove this box, blowing in a little smoke to drive the bees below. Let every bee-keeper see that his bees have "enough and to spare," remembering that they are faithful stewards and will return the trust with interest a hundred fold.

BEE PASTURAGE.

> " Bees work for man ; and yet they never bruise
> Their master's flower, but leave it, having done,
> As fair as ever, and as fit for use."

Propolis is a resinous gum, obtained by bees mainly from the leaves, buds, and trunks of plants and trees, and used chiefly to fill up the holes, and plaster the inner surfaces of the hive.

Bee-bread, called also pollen and farina, consists of the fecundating dust of blossoms, and is also collected by the bees from the fine dust of flour—rye flour is best—and constitutes especially the food of their young.

Honey is the great staple of the bee-hive. It varies in quality and value, according to the source from which it is obtained. The three principal honey harvests with us are, 1. From the blossoms of fruit trees in spring ; 2. and greatest of all, From white clover ; 3. Buckwheat. The honey locust, the basswood, the whitewood, oak, maple, and other forest trees, and the flowers of a great variety of plants, also yield large quantities of honey. Our *surplus honey* should, if possible, be secured from the white clover, it being of the best quality. This is also the period of the most rapid comb building, empty hives being sometimes filled with comb in six days. Bees will often gather sufficient honey for their winter stores from buckwheat alone, when it is near by, the season favorable, and they have plenty of empty comb in which to store it. Mr. Alvin Wilcox, of West Bloomfield, N. Y., is said to have had two swarms increase in weight twenty pounds from buckwheat in a single day. The field was within fifteen rods of the apiary. "The Baron of Berlepsch has had single colonies in his apiary which increased eleven pounds in weight in one day. Mr. Kader, of Mayence, had one which increased twenty-one pounds, and the Rev. Mr. Stein, of the same place, one which increased twenty-eight pounds, in a day."—*Bee Journal* for July, 1861, *p* 164.*

*I have known a small second swarm, in the honey season, to store sufficient for winter in ten days, when empty combs were provided.—*M. Quinby*

RANGE OF PASTURE.—OVERSTOCKING.

Bees will bring stores from a distance of three or four miles ; but the nearer they find supplies, the more rapidly they accumulate them ; and they will often accept an inferior quality, rather than go a great distance for a superior. So great is the yield of good pasturage, that there is but little danger of overstocking. From a report made to the Austrian government, on the state of bee culture in twenty-one states of that Empire, furnished in an excellent article on " Bee Culture," by Mr. Bruckisch, of Texas, in the U. S. Patent Office Report for 1860, *p.* 282, we quote as follows :

" Average number of bee-hives to one square mile :

Transylvania	300	Serbia Banat	400
Croatia	320	Carinthia	500
Gorz	340	Styria	510
Galacia	350	Carniola	900
Lombardy	360		

In Mr. Langstroth's valuable work on the *Honey Bee*, we find statements from Mr. Samuel Wagner, Editor of the *Bee Journal*, Philadelphia, in which he says, that the present opinion of the correspondents of the German *Bee Journal* appears to be that a district cannot readily be overstocked, and Dzierzon says in practice at least, " *it is never done.*" In Russia and Hungary, apiaries numbering from 2,000 to 5,000 colonies are frequent ; and 4,000 hives are sometimes congregated, in autumn, on the heaths in Germany. Oettl says :—

" When a large flock of sheep is grazing on a limited area, there may soon be a deficiency of pasture. But this cannot be asserted of bees, as a good honey-district cannot readily be overstocked with them. To-day, when the air is moist and warm, the plants may yield a superabundance of nectar ; while to-morrow, being cold and wet, there may be a total want of it. When there is sufficient heat and moisture, the saccharine juices of plants will readily fill the nectaries, and will be quickly replenished when carried off by the bees. Every cold night checks the flow of honey, and every clear, warm day reöpens the fountain. *The flowers expanded to-day must be visited while open ; for, if left to wither, their stores are lost.* The same re-marks will apply substantially in the case of honey-dews. Hence, bees cannot, as many suppose, collect to-morrow what is left ungathered to-day, as sheep may graze here-after on the pasturage they do not need now. Strong colonies and large Apiaries are in a position to collect ample stores when forage suddenly abounds, while, by patient, persevering industry, they may still gather a sufficiency and even a surplus, when the supply is small, but more regular and protracted."

Localities differ as widely in their resources of honey as in pasturage for cattle ; and the yield of any particu-lar locality is very much affected by the season : hence it is impossible to say how many stocks can be sustained to the square mile. Very few places in our country are, as yet, in any danger of being overstocked.

PROFITS OF BEE CULTURE.

" A penny saved, is two pence earned."—*Poor Richard.*

The profits resulting from bee-keeping, depend mainly upon the locality and season—presuming, of course, that the bees are well taken care of.

The bee editor of the *Rural New Yorker*, (in No. for Jan. 21, 1860,) says, " We are satisfied that nothing will pay better than keeping bees ; but care is required, and a knowledge of their habits, and, for want of this, many fail."

The same number contains the Bee account of Mr. Hiram W. Bulkley, of Saratoga co., for 1859, reported to the State Agricultural Society, as follows :

1859. *Dr.*

June 1.—To 29 swarms wintered, worth $7 each	$203	00
30 hives for new swarms, $1,50	45	00
100 honey boxes, 18c	18	.00
13 frames on which hives are suspended, 50c.	6	50
labor bestowed, estimated at	10	00
expenses of marketing	3	46
	$285	96

Cr.

By sales of 520½ lbs. clover honey, including boxes, 25c.	$130	12
By 489 lbs. buckwheat honey, including boxes, 14c	68	46
By seven swarms taken up, honey estimated at $3 each	21	00
By 7 hives for use again, $1,50	10	50
By honey on hand and used in family, estimated	30	00
By 35 swarms on hand, $6	210	00
By 17 swarms sold after honey season	100	00
By First premium at State Fair	5	00
By First premium at Saratoga County Fair	1	00
	$576	08
Deduct debits	285	96
Profits	$290	12

In the same paper for March 3, 1860, Mr. T. S. Underhill, of St. Johnsville, N. Y., writes :

" The amount to be obtained from any particular section, of course depends on its fertility and the sources of honey. Mr. A. W. Ford, of Middleville, Herkimer Co., N. Y., has, the past season, from an apiary of 130 stocks, received an increase of 170 swarms, and 6,000 lbs. surplus honey, which sold at 20 cents, and the swarms at

$4, making an income of about $1,800 from a capital of
$600 or $700. This, he says, is better than he has ever
done before, but it shows what may be done in a good
locality, with a favorable season. M. Quinby, of St.
Johnsville, N. Y., who has, I believe, the most extensive
apiaries in this country, and is a practical bee-keeper,
says, in his treatise on bee-keeping : 'In some seasons,
particularly favorable, your stocks collectively will yield
from one to two hundred per cent. I have known a single stock in one season to produce more than twenty dollars in swarms and honey, and ninety stocks to produce
over nine hundred dollars.' He speaks of these as instances of an extra yield, and further remarks, ' that a proper
estimate can be made only by the average of proceeds of
several years;' but that ' a single stock, rightly managed,
in the long run, is worth more than $100 at interest.'"

Mr. R. H. Davis, of Larone, Somerset county, Me., is
reported in the Maine *Farmer*, as having received from
four swarms a clear profit of $67 25, that is, a net of over
three hundred per cent.

But California is the Paradise of bees. Mr. Hamilton,
of Stockton, reports in the *Sacramento Union* for Jan. 14,
1861, that he had, the previous season, thirty-five swarms
of bees, which increased to five hundred, and the yield of
honey for the season was 20,075 pounds, making an average from the thirty-five original swarms of 573 pounds
each ! Mr. H. moved his bees Feb. 1, 1860, from Stockton to Santa Clara, where they remained till July 1st,
when the swarms had increased to 270. He then returned
with them to Stockton, and by the first of October the
swarms had increased to 500.

In 1860 bees were worth in California $25 a swarm, and
honey 50 cents a pound. With these figures, the reader

may make his own estimate of the profits of bee culture in California. Incredible as the figures appear, other parties report results nearly as great.

In view of the facts given—and they agree with the experience of intelligent bee-keepers every where—it appears safe to estimate the *net* profits of keeping bees *at least* at one hundred per cent. per annum, when they receive the attention and care that a farmer gives to any other kind of stock. Even at this low estimate, can the farmer give his attention to any thing else that will pay as well?

Considering the small capital needed to begin with ; the ease with which it may be expanded, and the safety of the business—with the right kind of hive—the lightness of the labor, and little attention needed ; the very small waste from "wear and tear," and that what the bees gather is so much actually saved—bee-keeping commends itself to every producer whose situation will admit of it. Every family in the country, and many families in cities, might keep a few swarms of bees ; and thus, if they did not sell any honey, they would add a material item to their own tables. Millions of dollars are lost in our country every year from want of bees to save it.

HUNTING AND HIVING WILD BEES

I will give the method practised by myself, whereby I have no difficulty in soon determining the exact locality of the swarm, and securing it.

As something depends upon the *season of the year* in which it is proposed to hunt them, I will give the different methods suitable for each season, beginning with early spring. Take the middle of a warm sunny day, the thermometer about 48° in the shade ; go to the woods near the supposed locality of the wild swarm, and with a lighted match or candle, burn a little dry honey-comb, beeswax, or piece of wood, on which a few drops of oil of anise have been poured. Keep a gentle " smudge" (to use a bee-hunter's phrase) going for 15 to 30 minutes, or until the bees come searching along close to the ground, following the line of the smoke. A foot or two from the " smudge," in the direction in which the wind is blowing, elevated a foot or two from the ground, if the surface be smooth—if bushy, higher, so as to have it the highest object near the smoke—place a piece of honey-comb, partially filled with sweet, freshly diluted honey ; or sweetened water will do if the swarm is close by, otherwise they are not so sure to readily accept it. If a drop or two of the oil of anise be added to it, or sprinkled on the comb, the bees will be attracted by its strong scent, and work all the more rapidly. The bees will soon begin to collect upon the comb, and if the weather continues favorable, with but little wind, and the swarm is near by (by which I mean within half a mile), a steady line of bees will be seen going from the combs, laden with the sweets, to their home, wherever that may be.

The first time a bee starts for home, and sometimes for several of the first trips, it will be seen to describe a circle immediately around the comb, the circles gradually becoming larger and larger, till apparently the true bearing is found, when a "bee line" is struck for home. In order to watch their course as far as possible, an open space must be chosen, or what is better, an open field, even if it be somewhat further off, when we shall be able, by keeping the eye as close to the ground as possible while the bees fly against the sky for a back-ground, unobstructed by trees or other objects, to more perfectly line them. Of course new comers are constantly arriving, and these latter will fill the air with their spiral curvings ; but a little practice will enable the hunter's eye to catch those whose flight will now be straight for home, without more than a part of a single circle, while even those striking spirals (if it be not windy), will evidently lean toward home, or circle from the combs in that direction.

Many different swarms are often thus set at work from the same spot at once, sometimes causing much vexation. This will at once be known by *constant quarreling.* If, now, we desire to *divide* them and get rid of all but those which go in a direction indicating that they are probably wild, we have only to place a box supplied with clean honey-comb, and a little honey in the cells, in the spot from which the bees have been working, removing all other comb, and after the bees have collected therein, close them in with a cover, and carry them in the direction where we suppose the swarm to be, and as nearly to the spot as we can guess. If we diverge a little out of the line, and yet, when we set them again at work, be considerably *nearer* the swarm we are in quest of, while we are *farther off* from the others, we shall pretty effectually divide them. But we must not move too far at a time, for if we chance to go beyond the tree, our bees will not be likely to return. If we have chosen the right spot, we shall now probably get a "cross-line," and by following *both lines* accurately to *the point where the one crosses the other,* we shall be in their immediate neighborhood. If left to themselves, one swarm only will be at work after two or three hours, usually the *nearest* and *hungriest* swarm driving the others off. Hence not the best, but the poorest and least valuable, is often found if we do not divide them.

Since it often happens that a hungry and more distant

swarm will work more rapidly than one having a good
supply of stores on hand that is nearer by, and as we
cannot always determine whether the swarm be located
in the adjacent woods or at a hive beyond, by *guessing*, the
distance may be very nearly found in the following manner :

Get the bees at work at two points a little distance
from each other, and with suitable instruments construct
a triangle, making the distance between these two points
its base, as *A B* in the annexed diagram : the bees
diverging from the line *A B* in the directions *A D*, and
B C, respectively.

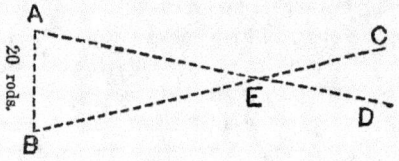

Fig. 29.

If we have constructed the triangle correctly, the distance between the points *A* and *B* is proportional to the
distances between *E*, where the lines *A D* and *B C* cross
each other, and *A* and *B* respectively. If, now, the distance
between the two points where we have the bees at work,
be 20 rods, and the length of the line *A B*, in our triangle, be made 20 inches, as many inches as there are in
the line *A E* or *B E*, the location of the swarm is distant
from the points *A* and *B*, in rods.

When their vicinity is found, as indicated by the point
where the lines cross each other, we must carefully mark
the place and commence searching for the tree. This is
the most tedious of the whole process, often requiring the
nicest skill *in getting into the right position* to discover bees

at altitudes in which they are often found. When the
trees are short and small, it is not a difficult matter to
see them ; but when the bees are 40 to 60 feet from the
ground, it is another thing altogether. In any case,
the way to find them is, when you have nothing to aid
the naked eye, to get into the shadow of the tree, and
walk slowly backward and forward so as to bring every
point of its body and larger branches in range between
the eye and the sun, looking at the sides of the tree just
below the sun and outwardly, carefully and slowly. The
bees will be seen very easily while in this position, and appear quite large from the reflection of the sun's rays striking upon their wings. A good spy-glass is a great help,
however, and by its aid one can readily determine whether
bees are working in and out of a tree or not, even by looking over the top and sides of the branches, or through
openings almost anywhere about the tree.

Bees will work honey at any time, even in mid-sum-
mer, if it be fresh from the hive. The way to set them at
work in the summer season is, with a cup or box, with a
cover, catch one at a time from the blossoms, set the cup
on the stump of a tree, or other convenient point, till no
humming is heard in the cup, when the cover is very
carefully removed, and the bee allowed to get its fill of
honey undisturbed. It is usual with bee-hunters to make
a bee-box for hunting purposes, with a slide two inches
from its bottom, so that the comb and honey may be shut
out from the bees while catching and carrying them, to
prevent their becoming besmeared with honey ; for when
they do ever so little, there is no use trying to do anything with them, for they know as well as you can tell
them, that honey is of no consequence so long as they can
not get home with it. A bee-tree should never be cut,

except by a person of experience, before the middle of May or after the first of September, since it will be extremely difficult to save the bees at such times. During these months no trouble need be feared by any person. Simply remove the honey and comb, after subduing the bees a little with smoke of old rags or tobacco, and with a stick or nail fasten a small piece of the comb, containing eggs and brood only, in a box, for a temporary hive. Now hive the bees precisely as you would a young swarm in swarming time. Remove them at night to your domicil, having previously placed the brood combs as evenly and carefully in a hive as it is possible for you to do, transferring thither the bees, and giving them but little or no honey. If movable frame hives are at hand, this is easily done. Should the queen have been destroyed, they will soon rear another, and collect far more honey, and prove a thousand times more satisfactory to you, than if you had waited till fall, and then cruelly destroyed them all, as is so often done for their stores alone. I have no patience with that class of bee-hunters, or bee-keepers who practice this latter barbarity. It is too much like the practice of the rude Indians of the forest, who annually slay whole herds of deer for their skins alone. A more perfect parallel still would be found in the farmer who should make yearly slaughter of his beeves and other stock for their hides, throwing their carcasses to the winds !

After a bee tree is cut and the bees "broken up," robbers from neighboring trees or otherwise soon make their appearance, appropriating the spoils. If any wild swarms are in the vicinity, as is most always the case if the tree contains an old one, they are easily followed home and captured. I have often found three or four in a circuit of a hundred rods.

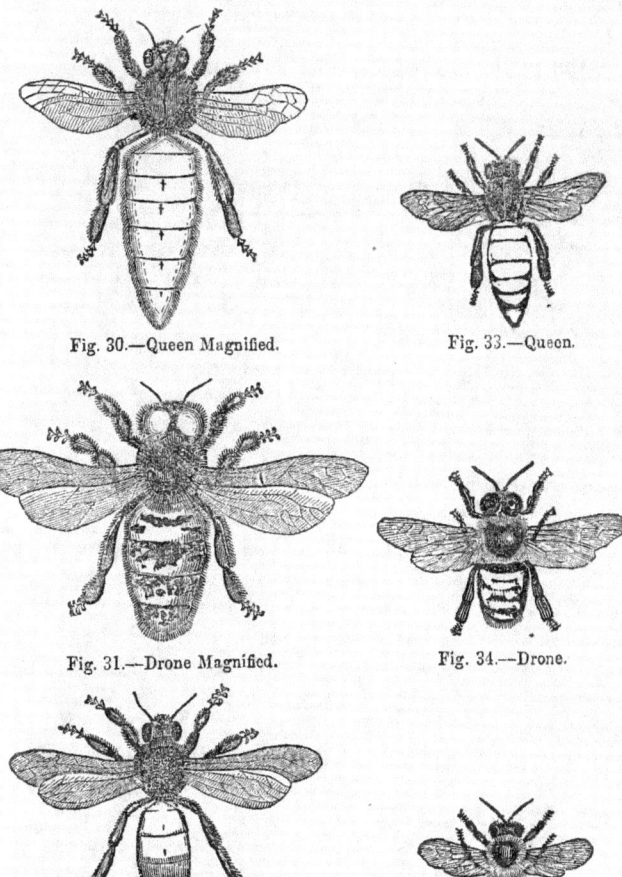

Fig. 30.—Queen Magnified.

Fig. 33.—Queen.

Fig. 31.—Drone Magnified.

Fig. 34.—Drone.

Fig. 32.—Worker Magnified.

Fig 35.—Worker.

ITALIAN BEES.

The Italian Bee, called also the Ligurian Bee, is found in a small Alps-pent district, embracing portions of northern Italy and southern Switzerland. They were once more largely distributed ; and are described by Aristotle, Virgil, and other ancient writers, as "small and round in size and shape, and variegated in color," and the most valuable of any then known. Various accounts of bees answering the above description, as once known but perhaps lost, have reached our day, but had come to be regarded, for the most part, by sober matter-of-fact moderns as among the fictions of ancient mythology. But it is now believed that these wonderful little creatures, thus thought worthy of preservation in song by one of the world's greatest poets, still exist and are identical with those now called Italian bees. As they were described two thousand years ago, so they are found now, the most valuable and industrious of their kind. Why they should have been lost sight of for so many years, does not appear, unless it be in consequence of that universal, well known law of nature by which the inferior type predominates over the superior, if neglected ; the golden-hued bee being thus gradually displaced by its black rival, except (so far as is known) in the district named, where the superior race appears to have held exclusive possession, the surrounding mountains, covered with perpetual snow, being impassable by their wings.

They were accidentally discovered by Capt. Baldenstein, while stationed in Northern Italy in the wars of Napoleon ; who after returning to his castle in Switzerland, procured, in 1843, a colony near Lake Como, and transported them over the Alps to his northern home. They were introduced into Germany by Mr. Dzierzon in 1853, and soon became very popular.

The first successful importations into this country were made in 1860, by Messrs. S. B. Parsons, of Flushing, L. I., P. J. Mahan, of Philadelphia, and R. Colvin, of Baltimore. In 1861, Mr. C. W. Rose, of New York city, succeeded in bringing six colonies to this country, out of forty-nine purchased in their native district. I believe one or two other importations have been made. The Italians have already been extensively introduced into the apiaries of both the Atlantic and Pacific sections of the United States, and are becoming great favorites.

If I am rightly informed, the importations of Messrs. Mahan and Colvin were from Germany ; while those of Messrs. Parsons and Rose were direct from Italy, the latter under the personal care of Mr. August Bodmer, an experienced Tyrolean bee-keeper, who selected them in their native fastness and brought them hither.

It is claimed by each of these importers, that he has *the genuine Italian bee.* Whether there is any choice or different degrees of purity in these stocks, I do not pretend to decide, but do know that *very great care is requisite to breed them pure,* and the purchaser cannot be too careful of whom he obtains his queens. I have examined samples of most, I believe all, of these importations, have queens bred from two of them, and am *not yet* satisfied as to what constitutes the distinctive characteristics of the Italian race. If color be the test, I am still lost ; for I have failed to find any that are constant in this respect, or in the number and breadth of the yellow bands on their abdominal rings. So far as my experiments with them go—and they have now been extended through a period of two years—they indicate that the lightest colored queens produce the most

brilliant drones ; and if the drones are the offspring purely
of the queens, as is generally admitted, may it not be in-
ferred that *possibly none are pure*, but that all have a trace
of black blood in them, showing itself most strikingly in
queens and drones, which must be carefully bred out before we can determine their real characteristics and render them uniform in shape, and constant in color ?

The illustrations (magnified) on page 75, were engraved
from selected specimens, and show the distinctive markings and bands in great perfection—the yellow portions
being represented by white. It will be observed that the
body is more round and slender, and a little longer, the
wings somewhat larger in proportion to the body—in fact
the whole bee appearing more delicate in form and tex-
ture—than the common bee.

The testimony of several well known apiarians who
have had experience with the Italians, is here given, that
the reader may the better judge whether this new variety
is worthy of his attention.

From the Baron of Berlepsch and Mr. Dzierzon :

The Baron of Berlepsch and Mr. Dzierzon, among the
most intelligent and skillful bee-keepers of Germany, give
the Italian decided preference over the common bee. The
former says he has found—

" 1. That the Italian bees are less sensitive to cold than
the common kind. 2. That their queens are more prolific. 3. That the colonies swarm earlier and more frequently, though of this he has less experience than
Dzierzon. 4. That they are less apt to sting. Not only
are they less apt, but scarcely are they inclined to
sting, though they will do so if intentionally annoyed
or irritated. 5. They are more industrious. Of this fact

he had but one summer's experience, but all the results and indications go to confirm Dzierzon's statements, and satisfy him of the superiority of this kind *in every point of view*. 6. That they are more disposed to rob than common bees, and more courageous and active in self-defence. They strive on all hands to force their way into colonies of common bees; but when strange bees attack their hives, they fight with great fierceness, and with an incredible adroitness."

From Mr. F. A. Deus, and others :

Mr. F. A. Deus, who with three other members of the German Apiarian Convention, held at Mayence, in 1856, in that year made a tour of France, Switzerland, and Italy, in search for and observations on Italian bees, in his report, as quoted in the *American Bee Journal* for Sept., 1861, *p.* 213, says :

" At the Villa di Negro, near the latter city (Genoa), the genuine Italian bee exists in all its beauty and perfection. It was delightful to observe the celerity, agility and grace displayed in all their motions by the busy workers, as they rifled the flowers of their sweets. Their bodies were so slender and delicate, their colors so bright, and their markings so clear and distinct, as to surpass greatly any specimens of the race which had previously come under our notice. We caught a number of them, and preserved them in alcohol for future comparison. * * * It is evident that the Italian bee is not a mere climatic variety, but really a distinct race. We were repeatedly assured also that the common kind only was found in the Kingdom of Naples and in the warmer districts of Upper Italy. We chanced to fall in with a bee-keeper from Normandy, who informed us that two kinds of bees were cultivated in that country—the common kind and also a yellowish or orange variety. The latter, he stated, were much preferred, as being more gentle and more industrious. The common kind, he said, were particularly irascible and wild. This account likewise corroborates the opinion that the Italian bee is not the common insect modified by special climatic influences, because Normandy differs little in that respect from Central Germany.

"At Lago Maggiore and Lago di Como, we found Italian bees exclusively, and of the most perfect type, like those of Genoa. These districts, indeed, appear to be their chief *habitat.*"

From the American Agriculturist :

"The fact that so many of our oldest apiarians have considerable confidence in them, argues well in their favor. We have watched their multiplication from a single swarm, and if the rate of increase be as great at other points to which the queens are being daily dispatched, it will not take long to fill the country with them—if such a consummation be desirable. Below we give an extract from a letter, dated August 10, written by Mr. E. A. Brackett, the well known sculptor, who is an enthusiastic amateur in bees also. His suggestion in regard to improving bees, by care in selecting breeding queens, is worthy of attention. All kinds of domestic animals have been brought to a much higher standard, by special care in breeding. Why may not our common bees be in like manner improved ?

"'My experience thus far, satisfies me that they have

not been overrated. The queens are larger and more prolific. The workers, when bred in comb of their own building, are longer and their honey sacs larger. They are less sensitive to cold, and more industrious.

" 'In all my handling of them—and I have done so pretty freely, lifting the combs, and examining them almost daily —I have never known one to offer to sting. A queen that I received in June, and introduced to a strong stock of bees, in eleven days filled thirteen sheets of comb with brood and eggs. There is at present scarcely a black bee in the hive, so rapid has been the change. Although I have taken from it large quantities of worker brood and sealed drones, the hive is still overflowing.' "

From the California Culturist :

" We believe, however, that the superiority of the Italian bee is no longer questionable, even among apiarians who have large stocks of the common bee for sale. We take pleasure in introducing proof of this, that those who may have been in doubt may have their doubts removed, and at once obtain this superior breed ; just as the stock-grower would a superior breed of horses, cattle, or sheep."

From Mr. L. L. Langstroth :

" It is hardly necessary to say, that a species of the honey-bee so much more productive than the common kind, and so much less sensitive to cold, will be of very great value to all sections of our country. Its superior docility would make it worthy of high regard, even if in other respects it had no peculiar merits. Its introduction into this country will, it is confidently believed, constitute a new era in bee-keeping, and impart an interest to its pursuit which will enable us, ere long, to vie with any part of the world in the production of honey."

A year later, (Aug. 24, 1860,) Mr. L. wrote to Mr. Parsons
as follows :

"I have three colonies (artificial swarms) to which
Italian queens were given in June. All of the common
bees appear to have died ; and if we may judge from the
working of these colonies, the Italians will fully sustain
their European reputation. They have gathered more
than twice as much honey as the swarms of the common
bee. This, however, has been chiefly gathered within the
last few weeks ; during which time, the swarms of common bees have increased but very little in weight. The season has been eminently unfavorable for the new swarms,
(one of the very worst I ever knew) and the prospect is,
that I shall have to feed all of them except the Italians.

L. L. LANGSTROTH."

From J. P. Kirtland, Cleveland, Ohio, Sept. 13, 1860 :

"*First.*—Their disposition to labor far excels that of the
common kind. From the earliest dawn of day to the ar-
rival of evening, they are invariably passing in and out
of the hive, and rarely suspend their work for winds, heat
or moderate showers—at times when not a solitary individual of the common kind is to be seen. Two hours each
day, their labors are extended beyond the working time
of the last named kind.

"*Second.*—Power of endurance, and especially of resisting
the impression of cold, they possess in a marked degree.
Since the buckwheat, salidagoes, and astors have flowered
in this vicinity, the nights have been remarkably cold.
This low temperature has in a great measure suspended
the efforts of the common bees, and they have been eating
their previously accumulated stores. Not so with the
Italians ; they have been steadily accumulating honey

and bee-bread, and rapidly multiplying their numbers They seem peculiarly adapted to resist the chilly atmosphere and high winds, which predominate in autumn, on the shores of Lake Erie.

" *Third.*—Prolificness they equally excel in. Both my full and half-blooded stocks have become numerous and strong in numbers, as well as in stores, at this late season of the year, when the common kind have ceased increasing, and have become nearly passive.

" *Fourth.*—Their individual strength is greater ; and this is well illustrated in their prompt manner of tossing to a great distance any robber that chances to approach their hive.

" *Fifth.*—Their beauty of color and graceful form render them an object of interest to every person of taste. My colonies are daily watched and admired by many visitors.

" *Sixth.*—Of their moral character, I cannot speak favorably. If robbery of weaker colonies is going on, these yellow-jackets are sure to be on hand. So far as my experience has gone with them, I find every statement in regard to their superiority sustained.

" They will no doubt prove a valuable acquisition to localities of high altitudes ; and will be peculiarly adapted to the climate of Washington Territory, Oregon, and the mountainous regions of California.

J. P. KIRTLAND."

From the American Apiarian Convention :

The following from the report of the American Apiarian Convention, held at Cleveland, Ohio, March 12–14, 1862, is valuable in their favor, after a trial of three years :

" *Italian Bees.*—All agreed as to the superiority of the

Italian to the common black bee. They deserve all the good things that European bee-keepers had said of them, save one. They are not more peaceable, but more *irascible* than the black bee, and their sting is more poisonous. Mr. Langstroth gave it as his experience, and that of some of his friends, that the Italian bees, instead of being more peaceable than our common kind, are more irascible (except in the season of honey gathering), and are more difficult to quiet when once excited. The Italian who brought all Mr. Parsons' bees, said that our bees were far more peaceable than the black bees of Germany. A German writer who furnished a valuable article on bee-keeping, for the Patent Office Report of 1860, says that our bees are much more easily handled than those of Germany. This accounts for the belief in Germany, that the Italian bees are more peaceable than the black species. The remarks of Prof. Kirtland seemed to sum up all that other gentlemen had said of the Italian bee. The professor prefaced his remarks by saying that he had no " ax to grind," and no bees to sell, and would not have until his experiments had been completed, which would be three or more years. After discussing the good qualities of the Italian bee, he said that it was as much superior to the black bee as Shorthorn cows and Chester hogs are to the "scrubs" of the country ; and that the Italian bee is : — 1. Stronger, more active, and resists lake winds and chills better than the common bee. 2. It works more hours every day. 3. It collects more stores. 4. It works on some flowers which the black bee cannot operate on. 5. It breeds more freely. 6. It is more irritable, and its sting more painful. 7. It is more beautiful. 8. It, in short, compares with the common bee as the Short-horn

Durham does to the native scrub. The Dr. cautioned against breeding " in and in," and he and other gentlemen advised bee-keepers to purchase queens both from the Parsons and Rose stock, to prevent too close breeding."

My own experience with Italian bees, differing somewhat from those above given, has not yet been such as to warrant a decided opinion in all respects. Of their general superiority, however, there can be no question. I have found them quite as gentle as the common kind ; for, though quicker on the wing, I have been stung only twice by them, so as to cause any swelling or pain, during a daily and almost hourly handling of them for two years. I have, moreover, been too much engaged in queen rearing and experiments to determine defin tely their comparative industry. They seem to fly swifter and work more hours than common bees, which they easily master, and whose stores they appropriate. I noticed, the past season, that one of my Italian hives was rapidly accumulating honey, while others were diminishing in weight. Looking into the hive, a short time after, I found it running over with bees, a large portion of them being black. I did not see them come nor know whence they came, as I had in previous instances of robbing ; but they were doubtless a subdued colony of blacks (not " contrabands"), which, after hard fighting, being spoiled of their treasures, had sought protection under the " ægis of the union."

HOW TO ITALIANIZE COMMON STOCKS.

The process of introducing the Italian race has been, to procure a queen, and after rendering a swarm of common bees hopelessly queenless by taking away its queen and successively destroying all queen cells for eight days, to introduce the Italian queen. It is found that the black bees will then accept her. And since the queen, once fertile, continues so for life, her progeny which will begin to appear in twenty-one days after her introduction, will become the exclusive possessors of the hive in a period of from three to six months.

So far the process is successful and perfect : but the *theory* has been, further, that inasmuch as the *drone* progeny of the queen was of necessity as pure as herself, whether she had met pure or impure drones, or, indeed, any drones at all, all we had to do to breed the Italian race in purity was, to get a pure queen, purely impregnated, and rear from her eggs a new race of queens for all our hives, when all our drones, being the offspring of pure Italian queens, would be pure Italians : then our queens, having been impregnated by black drones and hence producing hybrid workers, must be replaced with new ones reared from the original queen ; these new queens, meeting only pure drones, would insure the perfect Italianization of the entire apiary.

All this is theoretically plain enough, and it is presumed that not one person in ten will anticipate any difficulty in effecting a change of the apiary from one race to the other ; whereas, not one in a thousand, probably, will be able to do this on the first trial, if, indeed, at all. Let us look at the practical difficulties in the way of accomplishing it.

We will suppose that a pure Italian queen, purely impregnated, is obtained, and successfully transferred to a hive of common bees. After the lapse of twenty-four days, all progeny of the old queen will have disappeared from the combs. If we now exterminate the drones, we will insure Italian male bees in purity from the one hive only. But while rearing our *first* installment of queens for the purpose of procuring pure Italian drones for the impregnation of our *second* installment, very few *worker* bees of the Italian race are to be found in the apiary : hence the young queens, as well as others, *are nursed by black and hybrid bees*, producing in some way yet to be accounted for, a degeneration towards the black race.

The fact of such deterioration is admitted by those who have had the best of opportunities for judging, and can have no motive for misrepresentation. Although a few maintain that because dark colored queens sometimes produce as finely marked workers as the most brilliant ones, there is no degeneration ; forgetting that the characteristics of the workers are determined more by the impregnating drone than by the queen herself.

Prof. Kirtland, of Cleveland, Ohio, one of the closest observers, has entirely failed, we are informed, through two years of careful experiments, to produce a single pure queen from Italian brood transferred to hives of common bees.

Prof. E. Kirby, of Henrietta, N. Y., in attempting to account for this degeneration, suggests that the sperm or vitalizing fluid of the drone is perhaps supplied to the young larva while in process of transformation from a worker to a queen.

Mr. Langstroth, in *Bee Journal* for July, 1861, *p.* 166, although not admitting the point in issue, says :

" It seems very singular that the larvæ, which if developed as a worker would have been strongly colored, should in its transformation into a queen, lose all its brilliant yellow."

Again, while rearing queens by the removal of the reigning one, from *any* hive, there are more or less drone eggs deposited therein by *worker bees;* and if the latter are not purely Italian, we shall breed a race of *drones* of inferior quality, by which our queens may be impregnated, when, *theoretically,* we could not by any possibility have a black drone in the apiary.

Hence, it becomes evident, that if we hope to breed the Italian race in purity, we must establish a colony isolated from the black race, where they are to remain long enough to allow the progeny of the old queen to be displaced by the new. When the black bees have entirely disappeared, the queen may be removed, and the bees left to rear others. When capped over, the cells may be transferred to other hives, and the queen returned; repeating the operation when more are wanted. In establishing the Italian colony, it is believed to be of some importance that the Italian workers be allowed to renew the contents of the hive by filling it up with combs of their own construction; the cells built by them being somewhat larger than those built by the black race, and not having been used by the latter, may secure our queens from any possible taint.

In rearing Italian queens in great numbers, or indeed any other, it is advisable to establish small nuclei, or colonies of not more than one quart of bees to each. From such all impure drones may be easily destroyed; and the queens will mature and become fertile even sooner than from large swarms. But it should be borne in mind, that small colonies are more liable to the attacks of robber bees, and are more apt to accompany the queen in her flight to meet the drones. To prevent the latter, the presence of brood in the hive, in the earlier stages of development, should not be wanting. Such nuclei are perfect swarms in miniature : the prudent apiarian will keep a surplus on hand in summer to supply queens as wanted.

RENEWING QUEENS.

Queens gradually lose their fertility as they advance in age, producing fewer eggs and a greater proportion of drones. For this reason, after about the fourth year, the old ones should be destroyed and new ones substituted. In the recent Apiarian Convention at Cleveland, "Professor Kirtland said that after the third year the queen was nearly worthless, and should be killed, and a fertile queen put in her place instantly. So thought Mr. Langstroth: he said a vigorous fertile queen was worth half a swarm. Mr. Sturtevant thought the queen as good in the third year as at any time : and at four years he would not kill her, unless he *knew* that he could instantly get a young fertile queen in her stead ;—the risk was great, for at that season of the year the loss of a week or two was a serious loss."

"A fertile queen lays her eggs in regular order, commencing at a point and distributing them in circles, each surrounding the first, and on both sides exactly alike. . . Sealed worker brood should present a regular, smooth surface. An irregular brood denotes an unprolific queen."* A portion of raised oval cells in worker comb shows the presence of drone brood, and is objectionable, as indicative, except in the first laying of a young queen, of approaching barrenness.

TO PRESERVE HONEY COMBS.

They should be kept in a tightly closed box, and occasionally exposed to the fumes of burning brimstone to destroy all eggs of the moth. They are worth at any time during the summer almost their weight in gold.

* Bee Culture, pp. 162, 163.

PREVENTION OF SWARMING.

For about twenty days, in the swarming season, the bees may attempt to swarm. There is no way of confining the queen to the hive by contracting the entrance with "blocks," "gauges," or other *traps*, for the reason, among others, that bees vary greatly in size. Many fertile queens are able to go any where that a worker can, being longer but no larger. If our queens cannot fly, no *swarms* will be lost. But the queen may get down on the ground, in attempting to go with the swarm, and if a board be adjusted to the hive with one edge on the ground, she will be likely to crawl back into the hive, attracted by the great noise of the bees returning in search of her. If the attempt be seen, she should be found and returned. This rarely takes place, and only through neglect Even if *a queen* be occasionally lost, the swarm will rear another, or her place may be supplied from small nuclei kept on hand at this season for emergencies. Destroying, once in ten or twelve days, all queen cells, and giving the bees more room, will effectually prevent such attempts.

PURCHASING BEES.

Look into the hive to see that it contains a good stock of bees : they will show themselves at once, on being disturbed. The combs should be pretty regular, consisting of broad sheets of worker cells, and not small, irregular combs, or patches of drone cells which are worthless to transfer to movable frames. These things equal, the most valuable hive is the one whose contents will weigh the most.

MOVING BEES.

Bees may be moved to any distance at any time, but the summer is the best time. They should be shaded from the sun, kept as quiet as possible, and all jarring avoided. On a long journey, in hot weather, opportunities should be given them to fly daily, if convenient : the longer their confinement, the greater the liability of loss by disease. If in box hives, the hive should be inverted, its open end covered with gauze-wire cloth, and near its opposite end holes should be provided and covered with the same material for admission of fresh air. If in the Langstroth or similar hives, the combs must be first secured, or they may be broken down by the swinging of the frames, and the bees thus destroyed. Bees in my hives may be moved without preparation, except having the gauze-wire cloths in the honey board cleaned, and the front slide removed and a curtain of wire cloth substituted in its place. The combs are secure as in a box hive, for the frames cannot move.

A CHAPTER OF WELL SETTLED FACTS.

1. All stocks of bees should be kept strong in numbers.

A well garrisoned city may defy assault.

2. A moderate increase of swarms will keep them strong, and secure the largest yield of honey.

As the calves are raised at the cost of butter and cheese, so bees are multiplied at the expense of honey.

3. Bees filled with honey, are not inclined to sting.

As the robber's knife is stayed by your purse, so bees are bribed with proffered sweets.

4. In natural swarming, bees fill themselves with honey.

Emigrants to a new country carry their treasures along as capital to begin with.

5. Bees, alarmed with smoke or otherwise, instinctively seize upon their stores.

The householder at the cry of fire, secures what he can.

6. There should be no communication between occupied hives, allowing the bees of one to pass directly into the other.

" No house is large enough for two families."

7. A swarm of bees destitute of a queen fast dwindles away ; and, unless supplied with one, soon perishes either by robbers or moths.

A country without a government, a farm without an owner.

8. Swarms having combs insufficiently protected by bees, furnish a retreat for millers and food for worms.

Unguarded treasures invite thieves.

9. An excess of drones should be avoided by discouraging the construction of the cells that produce them.

Drones are the " dead heads" of the hive—the *useless males* in the farmer's herds.

10. The building of drone comb may, to a great extent, be prevented—first, by securing the construction of new combs in hives containing young queens ; and, second, by placing frames to be filled, in other hives, near the centre.

"An ounce of prevention is better than a pound of cure."

11. Queens are most economically reared in small swarms.

Who would employ ten men to do what one would do better ?

12. Small swarms, if united in the fall, will winter more safely, and consume less honey.

" In union there is strength."

13. Bees of colonies containing fertile and unfertile queens, should not be put together without first "breaking them up," i. e., inducing them to fill with honey, and destroying the unfertile queen.

14. *Natural swarming, always uncertain and perplexing, exposes the bee-keeper to much loss of time and money; while artificial swarming, securing at all times the presence of a worker-laying queen, doing away with all watching, and loss by flight to the woods, is both sure and economical.*

A CHAPTER OF SPECULATIONS.

The production of sex—has the queen any volition in it ? If the queen be watched during the early spring, while engaged in depositing eggs, she will be observed to pass by the drone cells ; but when the blossoms begin to yield honey in abundance, and bees become numerous, and disposed to swarm, the drone cells are prepared and eggs deposited therein. These latter eggs produce drones only, the kind of cell in which the egg is deposited *seeming* to determine, for the most part, the sex of the bee. It is found that eggs laid in drone cells produce drones only, while eggs transferred from drone to worker cells still produce drones. What determines the sex—the queen, the cell, or the nursing bees ? The prevalent *theory* has been, that the sperm sack of the queen, in the act of copulation with the drone, becomes filled with the sperm or vitalizing fluid of the male, without which all eggs possess sufficient vitality to germinate. Plausibility is given to this theory by the fact that the sperm sac of an unfecundated queen is found to be destitute of spermatozoa, while that of a fertile one is filled with them.

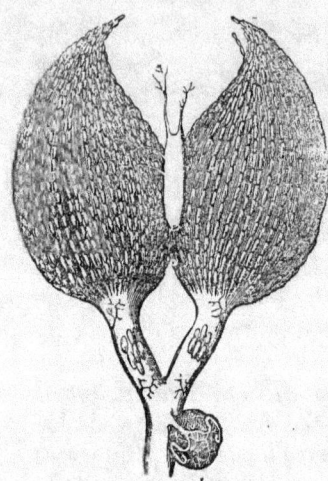

36.—Ovaries of queen.

Samuel Wagner, of Philadelphia, has suggested that in the act of depositing an egg in a worker cell the abdomen of the queen becomes compressed, forcing the sperm into contact with the egg in passing the sac, and changing it from a drone to a worker. [See Fig. 36.] But it is found that the queen will deposit eggs in cells but *just begun*, when no possible compression of the abdomen can take place, these eggs producing workers. Besides, it often happens that drones mature from the very smallest worker cells. It appears, after all, that not only the queen, but the workers also, know what the egg will produce ; for drone cells are not prepared until the season when they are wanted approaches, nor can the queen be induced to place eggs therein much in advance of this period.

For myself, I am satisfied that the *queen has* volition in the matter ; and that the bees have the power, by a particular process of nursing, suitable for each, to produce FROM WORKER EGGS EITHER QUEENS, WORKERS OR DRONES ! !

I am aware that this idea is somewhat new and ahead of "the books," but in my practice of rearing queens by very small nuclei, I have often found *drone brood maturing* in a small piece of worker comb cut from cards full of eggs taken from the centre of the hive, where it is scarcely possible drone eggs would be found.

INDEX.

ILLUSTRATIONS.

A BEE DRESS.

To the rim of a light summer hat sew a piece of gauze-wire cloth, cut six inches wide in the middle and tapering to two inches at each end, with curtain of any light cloth to tuck in about the neck.

But the BEST BEE DRESS is courage, kindness, steady movements, and a knowledge of the habits of bees.

"A soft answer turneth away wrath."

The advantages of this system are seen by the following:

I. WE LOSE NO SWARMS, *either during summer or winter.*

II. WE KEEP OUR BEES AT WORK, *and do not have entire swarms hanging to the hive, idle, throughout the honey-harvest.*

III. WE RAISE WHAT DRONES ARE NEEDED ONLY, *and do not feed a large number of supernumerary "dead-heads."*

IV. *All our swarms are, without interruption, provided with* FERTILE QUEEN.

Thus, by it,

1st. We have *complete control of our bees*, and are able to handle and take care of them, as easily as we do our chickens; consequently bee-keeping becomes a business as certain as any other.

2d. We do *not grope in the dark*, but have access to the interior of the hive, and can inspect the combs at all times, see their condition, remove them if necessary, or supply any want.

3d. By the use of a new, and entirely practicable style of comb-frames, we are enabled to exchange empty for full ones, giving the bees no excuse for lying idle; at the same time thereby increasing their numbers.

4th. When the stocks become sufficiently populous, *we rear new queens, and swarm them, just precisely as we would raise poultry, pigs, or sheep.* We don't wait for the "bees to act"—often on a single addled egg—while the season for that business is fast passing away. We thus do away with the necessity for watching them.

5th. The manner of producing new swarms, artificially, is new, easy, and rendered eminently successful by revolving the hive, giving the infant colony the strength and vigor of the parent stock.

6th. We make *all our new swarms really first swarms*, saving *ten days' time in breeding* over natural swarming; no eggs being laid in the combs of any young swarm *after the first for ten days.*

7th. The old and fertile queen remaining in the parent hive, honey gathering and breeding go on as before; while in natural swarming, honey gathering is nearly suspended for ten days, till a new queen is hatched, and no eggs are laid for twenty days, or until her fertilization.